1

Peter & Misty Phillip

UPSKILL

OR

DIE

Thriving
in the
AI Revolution

By Peter & Misty Phillip

Peter & Misty Phillip

UPSKILL OR DIE

Thriving in the AI Revolution

Spark Media

UPSKILL OR DIE

Thriving in the AI Revolution

Copyright @2025 by Peter Phillip & Misty Phillip

Published in association with Spark Media, LLC

https://MistyPhillip.com

Book Cover by Peter Phillip

Edited by Misty Phillip

https://www.trilogyworks.com

Print ISBN: 979-8-218-72238-8

Dedication

To our children, family, friends, and the countless innovators who inspire us daily. May you embrace the opportunities of the AI revolution with courage, curiosity, and a relentless drive to upskill and thrive.

To the readers who will hold this book, we offer it as both compass and companion. The artificial intelligence revolution is not a distant storm on the horizon. It is the dawn breaking around us right now. But within this disruption lies extraordinary opportunity for those willing to learn, adapt, and grow.

With love and gratitude,

Peter and Misty Phillip

Table of Contents

Part III: Action

Introduction

Peter stepped into the elevator with the weight of Enron's collapse still sinking in. On his final trek down the elevator, next to him was a woman in her sixties, her face wrecked with devastation, eyes glassy with shock. She was weeks from retirement, and with her life savings all invested in Enron stock, and got wiped out in a heartbeat. She now faced the grim reality of starting over, hunting for a job instead of looking forward to retirement.

Peter felt terrible for this poor woman. Watching the woman's devastation, a cold fear gripped him. But beneath the fear, a resolve hardened: he knew he must adapt. Unlike the woman standing next to him, who had been dreaming of retirement and now faced the complex, cold reality that she would have to start over. He knew he would be fine because he was intelligent, adaptable, and resilient. But this woman's reality deeply affected him. Knowing her life radically altered at a moment's notice. Life as she knew it would never be the same. Suddenly, everything changed!

While Enron's collapse was caused by fraud, not technology, the sudden, devastating impact on its employees offers a stark preview of what happens when the foundations of an industry crumble overnight. AI presents a potential scenario affecting millions, and it is coming faster than most people realize. The devastation Enron employees felt is like what it will be like for many people in the not-too-distant future. AI systems can now do many jobs faster, more accurately, and more cheaply. This will eliminate many jobs, and it will impact all aspects of our lives.

Millions of professionals will face the stark reality: artificial intelligence isn't coming for our jobs; it's already here, reshaping entire industries overnight. The question isn't whether AI will impact your career; it's whether you'll be ready when it does. Pandora's box has been opened, and tech companies are in a race for dominance, with no turning back.

Welcome to the age of "Upskill or Die."

That phrase might sound harsh, even alarmist, but it reflects the brutal truth of our current moment. We chose this title "Upskill or Die" to convey the seriousness and urgency of the situation: not to induce panic, but to help you prepare. This book is to serve as a call to action, prompting you to evaluate your current skills in light of the AI revolution. We stand on the precipice of the most significant economic transformation since the Industrial Revolution, and we want you to thrive.

Unlike previous technological shifts that unfolded slowly over decades, the AI revolution is happening at breakneck speed. ChatGPT reached 100 million users in just two months, and now AI-powered tools are automating tasks ranging from legal research to medical diagnosis, as well as financial analysis and creative writing. The pace of change is so rapid that the skills you mastered five years ago may be obsolete tomorrow.

However, here's what the doom-and-gloom headlines overlook: this unprecedented disruption also represents an unprecedented opportunity. Yes, AI will eliminate certain jobs, but it will also create entirely new categories of work we can barely imagine today. It will augment human capabilities in ways that make us more productive. To stay current, it's essential to learn how to use these tools.

This book is here to help prepare you, and we are your guides.

I'm Peter Phillip, an IT and emerging technologies professional with over 25 years of experience, and an expert in digital transformation. I've spearheaded innovative projects for Fortune 2 to 500 companies, and I'm a well-respected leader and trusted advisor. As a sought-after consultant, I've helped organizations navigate technological disruption by combining strategic foresight and practical expertise to foster sustainable growth.

I'm Misty Phillip, a visionary leader, entrepreneur, award-winning author and podcast host. With Peter's help, I founded and exited Spark Media after five years of creating learning opportunities and growth platforms for podcasters.

Together we co-founded Trilogyworks, a technology consulting firm specializing in advisory services for Cybersecurity, PQC, and AI-driven solutions for the enterprise.

We've worked together to blend technical insight and accessible wisdom in *Upskill or Die: Thriving in the Age of Artificial Intelligence.* This powerful, actionable blueprint enables confident navigation of the AI revolution and allows you to thrive within it.

We have observed technological trends and witnessed multiple waves of innovation reshape the world. We've seen companies rise and fall, careers flourish and falter, and individuals soar and struggle. These transitions failed or succeeded based on their ability to adapt to change. But we've seen nothing quite like what we're experiencing now. The AI revolution is distinct in both its speed and scope, affecting every industry, every job category, and every corner of human activity.

Over the years, one thing we've seen time and time again is that the people who thrive during periods of massive change share specific characteristics. People who thrive in their careers consistently embrace lifelong learning and continually develop skills that complement emerging technologies. Most importantly, they act rather than react, and they take control of their destiny rather than waiting for change to happen to them.

Upskill or Die is not a book about technology, though we'll explore AI's capabilities and limitations in detail. Upskill or Die is not a book about economics, though we'll examine how entire industries are being transformed. This is a book about you. The AI technological transformation will radically change the world. Upskill or Die will help you navigate this change.

Whether you're a mid-career professional wondering if your expertise remains relevant, a recent graduate entering a job market in flux, an entrepreneur seeking new opportunities, or someone nearing retirement questioning whether it's worth learning new tricks; this book is your roadmap to not just surviving but thriving in an AI-driven world.

The chapters ahead will equip you with three essential elements for success in the AI age: Knowledge (understanding what AI can and cannot do and its potential risks), Strategy (identifying which skills to develop and which to let go), and Action (practical frameworks for continuous learning and adaptation). We'll explore which jobs are most at risk and which are most secure. Identify the uniquely human skills that become more valuable as machines get smarter. We'll conclude with detailed, step-by-step upskilling guides for all levels and time commitments.

The AI revolution isn't something happening to you. It is an opportunity you can seize if you act now. The choice between upskilling and becoming obsolete isn't a choice between learning or not learning. You must choose between shaping your future or letting others shape it for you.

Transformation is available to you, but it requires action, and it requires action now! The window of opportunity won't remain open forever. Those who move quickly to develop AI-complementary skills will capture the best new jobs and opportunities. Those who wait will compete for an ever-shrinking pool of positions.

The future belongs to those who can work alongside artificial intelligence, not those who ignore it or wish it away. It's not a question of: Are you smart enough, young enough, or tech-savvy enough to thrive in this new world? The question is whether you're willing to do what it takes to upskill, adapt, and grow.

~ Your future self is waiting. Let's get to work.

Peter & Misty Phillip

Part I: Knowledge

Peter & Misty Phillip

Chapter 1
The AI Revolution Is Here

By the time this book comes out, it will already need to be updated. We are currently experiencing disruption at an unprecedented, accelerated pace; unlike anything we've ever seen before in human history. Ready or not, the AI revolution is upon us, and there is no turning back. We have opened Pandora's box, and we cannot close it.

The factory floors became eerily quiet after what had once been the heart of a bustling automotive plant in Detroit, where hundreds of workers had assembled car parts with practiced precision just a few years earlier. Now, sleek robotic arms moved in choreographed silence, guided by artificial intelligence systems that could detect defects invisible to the human eye, adjust for variations in materials in real time, and operate around the clock without breaks, sick days, or vacation time.

The scary part about these robots is that they are getting smarter every day. We can now teach robots in six hours what took six months to program two years ago. We aren't approaching an AI revolution; we are already in the middle of one. And most people haven't even realized it's begun.

The Stakes Have Never Been Higher

Artificial intelligence isn't just another technological advancement like the smartphone or the internet. It's the defining technology of our time precisely because it touches the one thing that has always belonged only to humans, intelligence itself. For the first time in history, we've created machines that can learn, reason, and decide. And they're getting better at it faster than anyone predicted. The problem is that no one knows how this will ultimately impact humanity.

19

Consider the breathtaking pace of recent developments. In 2022, most people had never heard of ChatGPT. By early 2023, it had become the fastest-growing application in internet history. But ChatGPT was just the opening act. AI systems can now write code, compose music, diagnose diseases, trade stocks, design products, and even create other AI systems to do multiple tasks. They're not just tools anymore; they're collaborators, competitors, and sometimes replacements.

The numbers tell the story. McKinsey estimates that generative AI alone could add $2.6 to $4.4 trillion annually to the global economy. Goldman Sachs projects AI could eventually automate 300 million jobs worldwide. The World Economic Forum predicts that 50% of all workers will need re-skilling by 2025, a prediction made before the current AI boom accelerated.

However, what makes this moment uniquely challenging is that previous technological revolutions gave us decades to adapt. The shift from agriculture to manufacturing unfolded over generations. Even the internet revolution took nearly two decades to transform how we work and live fully. AI is different, especially when you couple AI and robotics. Its capabilities are doubling not every few years but every few months or even every few weeks. The luxury of gradual adaptation no longer exists. It is time we get up to speed with what is happening in the AI revolution and figure out what part we can play.

Witnessing the Transformation

Over the past three years, we've watched the tremendous growth of AI. What we've discovered is both sobering and inspiring. The disruption is deeper and more widespread than most realize, but so are the opportunities for those willing to seize them.

In Silicon Valley, engineers who were initially concerned that AI coding tools would render their jobs obsolete have discovered these tools made them more productive than ever, handling routine coding tasks while freeing them to focus on architecture, strategy, and creative problem-solving. Their companies weren't laying off developers; instead, they were hiring more of them because AI-augmented teams could tackle bigger, more ambitious projects.

In every industry, AI is simultaneously eliminating some jobs and creating new ones. The people who thrive aren't necessarily the most technically skilled, the brightest, or even the youngest. They are the ones who embraced change earliest and most strategically. Creative thinkers who adapt well will thrive during this period of rapid AI development.

The Upskilling Imperative

The good news is that this reimagining reveals opportunities we never knew existed. As AI handles routine work, demand is growing for uniquely human skills, including emotional intelligence, creative problem-solving, ethical reasoning, complex communication, and strategic thinking. The businesses winning in the AI age aren't the ones replacing humans with machines. They are the ones combining human insight with artificial intelligence to achieve results neither could accomplish alone.

However, seizing these opportunities requires intentional action. The days of graduating from college, landing a career, and riding on the same skills for thirty years are over. In the AI age, learning isn't something you do just at the beginning of your career. It is something you do throughout your career.

The Promise of This Book

What you'll find in the pages ahead isn't another breathless account of AI's technical capabilities or dire warnings about technological unemployment. Instead, you'll discover a wake-up call accompanied by a practical playbook for thriving in an AI-transformed world.

We'll explore which skills may become more valuable as AI advances and those which may become obsolete. You'll learn frameworks for continuous learning that fit into your existing schedule and budget. We'll examine industries where AI creates more opportunities than it destroys and identify strategies for positioning yourself at the intersection of human creativity and artificial intelligence.

Most importantly, you'll develop what we call "AI fluency". Not the ability to program algorithms, but the wisdom to understand AI, where it excels, where it fails, and how to leverage its strengths while contributing to your uniquely human capabilities.

This book will show you how to become AI's collaborator rather than its competitor. You'll discover how to use AI tools to amplify your existing skills, identify emerging opportunities before they become apparent to everyone else, and build a career that becomes more valuable as artificial intelligence advances.

The Choice Is Binary

Across every industry, in every role, at every level, professionals face the same fundamental choice: grow or become irrelevant. This isn't hyperbole or scare tactics. It's the mathematical reality of exponential technological change. AI capabilities that seem impossible today will be commonplace tomorrow. Automation may eliminate jobs that feel secure this year. Skills that took decades to master may be surpassed by algorithms in months.

But here's the paradox that gives us tremendous optimism. The same technology that threatens to make some human skills obsolete also creates unprecedented opportunities for those willing to develop new ones. We're entering an age where human creativity, empathy, judgment, and wisdom become more valuable than ever, but only for those who combine these qualities with AI fluency.

The AI revolution is here, and the question isn't whether it will affect you. It already has, and it will continue. The question is whether you'll shape your response or let circumstances shape it for you. Your career, your financial security, your professional identity, and your ability to contribute meaningfully to the world all depend on how you answer that question. The cost of inaction has never been higher. But neither has the potential reward for those bold enough to act. The time for waiting is over. The time for upskilling is now!

Success in this new world requires immediate, strategic upskilling to combine uniquely human capabilities with artificial intelligence. But first, ask yourself this: when you look back on this moment five years from now, will you remember it as the time you took control of your future or the time you let the future happen to you? The choice is yours. And making the right one starts with truly understanding the technology that is forcing this decision upon us. In the next chapter, we'll demystify exactly what AI is and what it is not.

Peter & Misty Phillip

Chapter 2: What Is AI?

Picture this: you are chatting with your smartphone, asking it to find the closest coffee shop. In just seconds, it hears you, figures out where you are, sifts through a gazillion bits of info, and guides you with turn-by-turn directions. Not so long ago, this would've felt like something out of a sci-fi movie. Now, it's just another day in our lives. Beyond this, AI is already affecting professional fields, such as legal research and medical diagnostics.

Welcome to the wild, wonderful world of artificial intelligence!

Now, I know AI is all around us, but if you're like most people, you might still wonder, "Okay, but what is AI?" Some of you might imagine robots serving breakfast or supercomputers scheming to take over the world. The truth: it's both way more ordinary and way more incredible than all that.

At its core, AI is about teaching machines to perform tasks that we once thought only humans could accomplish. It's like giving computers the ability to recognize faces, understand our words, make informed choices, solve problems, or even create a piece of art that tugs at your heartstrings.

But here's the thing, AI doesn't think like you or me. It's not sitting there picturing a chessboard or getting butterflies before a big move. Take a chess-playing AI, for example. It's crunching millions of potential moves in a flash, picking the one that'll most likely lead to a win. It's a different approach, but it gets the job done with jaw-dropping smarts.

Understanding that AI operates fundamentally differently from human thinking is crucial for your strategic planning. You can't compete with AI by trying to think faster or process more data. Instead, you need to focus on uniquely human skills like contextual judgment, creative problem-solving, and relationship building. Your competitive advantage lies in understanding how to work with AI systems, not against them.

AI is the central force reshaping work and society as we know it today, and this necessitates upskilling to stay relevant. Let's begin by identifying key terms and definitions. So, what is AI? AI stands for Artificial Intelligence, currently the most talked about AI types are Large Language Models, computer systems or software that perform tasks typically requiring human intelligence, such as learning, problem-solving, decision-making, or pattern recognition.

What Are Large Language Models?

Large language models are AI that understands words. Large language models, or LLMs, are a special type of deep learning system designed to work with human language. These are the AI systems behind chatbots like ChatGPT, writing assistants, and many of the conversational AI tools people use today.

Developers have trained these models on enormous amounts of data, text, books, articles, websites, and conversations. These models learn not only individual words but also how words relate, how ideas connect, and how language functions in various contexts. It's like reading the entire internet and developing an intuitive sense of how people communicate.

What's remarkable is that LLMs don't just memorize the text they've seen before. They develop a kind of understanding of concepts and relationships, and can write about topics by combining ideas in new ways, answer questions by reasoning through problems, and even engage in creative tasks like writing stories or poetry.

The key insight is that language contains patterns that reveal how the world works. By learning these patterns deeply enough, an AI system can appear to understand topics it was never explicitly taught about. We can think of them as incredibly sophisticated computer programs that vast amounts of text from books, websites, articles, and other written materials have trained to learn the patterns, structures, and nuances of human language.

LLMs are your new research assistants, writing partners, and brainstorming collaborators. They can help you draft emails, analyze documents, generate ideas, and even code basic programs. The strategic question isn't whether you should use them, but how quickly you can integrate them into your daily workflow to multiply your productivity.

These LLMs are Generative AI, and this is what most people are familiar with and refers to AI systems that create new content, such as text, images, music, or videos, based on patterns learned from data. Many people think AI is just ChatGPT, but this is only one example of generative AI tools. It is impossible to list every generative AI program because of the sheer number and rapid evolution of the field. However, here is a current list of the most popular LLMs as of the writing of this book in mid-2025.

To give you an example of how quickly things are moving, let's look at the timeline for ChatGPT. ChatGPT launched on the scene with a demo in November 2022. ChatGPT is owned by OpenAI and is a versatile conversational AI for text generation, coding, and answering queries. GPT-3.5, with only text capabilities, was the original ChatGPT, and OpenAI retired this model in May 2025.

GPT-4 was a major upgrade with better reasoning, but it only had a life cycle of March 2023 to April 2025. Next up in July 2024 came GPT-4o mini, which was a smaller, faster, and cheaper version. The current flagship model is GPT-4o, and it handles text, images, and audio, and it's faster and cheaper than GPT-4. Special thinking models in the O-series are better at complex reasoning in math. It takes longer to respond, but it is more accurate.

As of the writing of this book, the most advanced and current model is ChatGPT 4.5, which launched in February 2025. OpenAI announced that GPT-5 will be released later in 2025, promising the next major breakthrough and significantly greater capabilities.

This rapid evolution cycle means that any AI skills you develop have an expiration date. The models you learn today will be obsolete within 18 months. Your upskilling strategy must focus on learning principles and workflows that transfer across AI generations, not just mastering specific tools.

Beyond ChatGPT

Here is a list of popular LLMs beyond ChatGPT.

Claude

Claude is an AI assistant created by Anthropic, a company focused on AI safety and research. Former OpenAI researchers, Dario Amodei (CEO) - former VP of Research at OpenAI and Daniela Amodei (President) - former VP of Operations at OpenAI, and several other AI safety researchers who previously worked on GPT models founded Anthropic in 2021. Anthropic's founders created the company specifically to focus on developing AI systems that are safe, beneficial, and understandable.

Gemini

Gemini is Google's flagship artificial intelligence model and the successor to their previous AI systems as LaMDA and PaLM. It represents Google's most advanced effort to compete directly with OpenAI's GPT models and other leading AI systems.

Grok

Grok is an AI chatbot developed by xAI, Elon Musk's artificial intelligence company. In late 2023, xAI launched Grok, a more rebellious and uncensored alternative to other AI assistants. This AI chatbot has a distinctive personality and real-time access to information from X (formerly Twitter). In July 2025, xAI launched its flagship model, Grok4, is it's positioned as a "PhD-level" AI that claims to outperform all other models. There are concerns about Grok 4 that extend beyond just another AI model release because this model brings us one step closer to ASI and AGI. More about that topic will be covered later in this chapter.

DeepSeek

DeepSeek is an artificial intelligence company and model series based in China that has gained significant attention for creating highly capable AI models at a fraction of the cost of Western competitors. The company has emerged as a major disruptor in the global AI landscape, particularly noted for its cost-effectiveness and impressive performance. However, because of our background in cybersecurity, we would never advocate using DeepSeek because it is an app created by the CCP. DeepSeek's lack of safety guardrails allows malicious actors to generate fully functional malware from scratch, including ransomware code, without requiring technical expertise. It also has a strong bias and fundamental security vulnerabilities in its implementation, suffers from poor data protection practices, and operates under a legal framework that enables state surveillance.

Your Strategic Takeaway

Different LLMs excel at different tasks. You may use Claude for in-depth research and analysis of complex client problems because it excels at nuanced reasoning. For rapid brainstorming and creative campaigns, use ChatGPT. When analyzing social media trends and sentiment, Grok's real-time Twitter access gives you current insights. By understanding each tool's strengths, you can increase productivity.

Don't be married to one AI tool. Investigate the strength and fluency of different LLMs, exploring their diverse capabilities to see what works best for you. This diversification protects you from vendor lock-in and maximizes your problem-solving capabilities.

AI Models, How They Work, and Why You Should Care

Artificial intelligence isn't just one thing. It's more like a toolbox filled with different smart systems, each designed to solve unique problems. Think of it like the difference between a hammer, screwdriver, and saw. They're all tools, but each one works best for specific tasks. These concepts are foundational to understanding the AI landscape and where it is going.

Rule-Based AI

The simplest type of AI works like a very sophisticated rulebook. These systems follow "if then" logic that programmers write out in advance. If a customer asks about store hours, then respond with the hours. If someone's credit card shows an unusual activity, then flag it for review.

This type of AI is everywhere, even though most people don't think of it as "real" AI. Your email spam filter uses rules to decide what's junk mail. Traffic lights use rules to control when they change. Even basic chatbots on websites follow pre-written decision trees to guide conversations.

The strength of rule-based AI is that it's predictable and explainable. You can trace exactly why it decided. The weakness is that it can only handle situations programmers thought of ahead of time. If something new comes up that doesn't fit the rules, the system gets confused.

Strategic Application

Rule-based AI is perfect for standardizing routine decisions in your work. If you make the same decisions repeatedly, consider using AI to streamline your workflow. Some examples are email prioritization, expense approval, and customer routing. By automating these types of tasks, it frees you up to do higher-level work.

Machine Learning

Instead of following a strict set of instructions, machine learning learns by looking at examples. It is kind of like teaching a child to spot a cat by showing them tons of cat pictures. Over time, they learn to recognize cats in new photos all on their own.

Here's how it works: You give the AI a bunch of data, like thousands of emails labeled "spam" or "not spam." It studies them, picking up on patterns. You know how terms like "free money" or "urgent" often show up in spam, or how emails from unknown senders seem fishy. The AI builds its own "rules" based on what it sees. Then, when a brand-new email pops up, it uses those patterns to guess whether it's spam. The more examples it gets, the smarter it becomes!

Machine learning comes in a few styles. Supervised learning is like having a teacher guide the AI with correct answers. Unsupervised learning lets the AI find patterns on its own, with no hints. And reinforcement learning is like training a puppy, rewarding good choices and nudging it away from bad ones.

Career Implications

Machine learning is your pattern-recognition superpower. ML tools enhance any job that involves analyzing data, spotting trends, or making predictions based on historical information. The key is learning to ask the right questions and interpret results, not building the models yourself.

Deep Learning

Deep learning is a type of machine learning that takes inspiration from how our brains work. Instead of humans telling it what to look for, it uses artificial neural networks — think of them as a simplified version of brain cells used to spot important patterns all by itself.

Picture these neural networks as layers of connected "neurons." Each layer picks up different details. For example, when looking at a photo, the first layer might notice simple things like shapes or colors. The next layer combines those into bigger clues, like edges or textures. Higher layers can even recognize whole objects, like a car or a face. It's like building a puzzle, piece by piece.

Deep learning figures out what matters with no hand-holding. Show it enough dog photos, and it'll learn that ears, tails, and furry patterns are key to spotting dogs. No one has to spell it out. This power drives some of today's biggest AI wins, like spotting diseases in medical scans, understanding your voice assistant, or translating languages on the fly.

Professional Opportunity

Deep learning is revolutionizing industries that work with complex data like images, audio, video, and text. If your field involves analyzing medical images, processing documents, understanding customer sentiment, or any task with rich, unstructured data, deep learning tools can become your competitive advantage.

Computer Vision

Computer vision is like giving computers a pair of super-smart eyes. It helps them "see" and understand images and videos, going way beyond simple pattern matching to making sense of the world like we do. These systems can spot objects, people, or animals in photos, read text in images, understand gestures, track movement in videos, and even pick up on facial expressions to guess emotions. Think of self-driving cars navigating roads, medical tools analyzing X-rays, or security cameras keeping an eye out for anything unusual, all thanks to computer vision.

How does it work? The AI breaks images into tiny bits, looking at patterns of light and color. By training on millions of labeled pictures, it learns to recognize everything from basic shapes to full-blown scenes with lots of action. It's like teaching a kid to spot his or her favorite toy in a busy room. The AI becomes better with practice.

Practical Applications

Computer vision is transforming every profession that deals with visual information. Real estate agents use it automatically to tag and categorize property photos. Doctors use it to spot anomalies in medical scans. Retail managers use it for inventory management and theft prevention. Quality control specialists use it to detect defects in manufacturing. The strategic question you should ask yourself is: what visual tasks in your job could be automated or enhanced?

Natural Language Processing

Natural Language Processing, or NLP, is all about helping computers understand and use human language with all its messy, wonderful complexity. It's not just about recognizing words but getting the meaning, tone, and even the sneaky sarcasm behind what we say.

With NLP, computers can translate languages, summarize big documents, answer questions about a text, or even chat with you like a friend. They tackle the wild world of human communication, including slang, cultural references, and those moments when we hint at something without saying it outright.

How do they do it? NLP mixes different tools: machine learning to figure out grammar, deep learning to catch meaning and context, and sometimes rule-based systems for tricky language rules. It's like teaching a computer to join our conversation, bridging the gap between our words and its digital brain.

Career Transformation

NLP is your secret weapon for any job involving text, communication, or language. Lawyers use it to analyze contracts and precedents. HR professionals use it to screen resumes and analyze employee feedback. Customer service teams use it to route inquiries and generate responses. Content creators use it to optimize for SEO and audience engagement. If you work with words, NLP tools can multiply your capabilities.

Reinforcement Learning

Reinforcement learning is like teaching a kid a new skill through trial and error, except it's AI figuring things out. Instead of showing it examples, you give the AI a goal and let it try different actions to see what works best. It's all about learning by doing.

Imagine teaching someone chess. They don't memorize every move. Instead, they play tons of games, learning what wins and what flops. Reinforcement learning AI does the same, but super-fast, playing millions of games in a flash to build smart strategies.

This approach has created AI that rocks complex games like Go or poker, controls robots for tricky tasks, or even optimizes things like cooling data centers or trading stocks. The AI discovers the best moves through experimentation, not by following a rulebook.

Strategic Thinking

Reinforcement learning is perfect for optimization challenges in your work. It excels at resource allocation, scheduling, pricing strategies, and any scenario where you need to balance multiple competing objectives. The key insight is that RL can discover counter-intuitive solutions that human experts might miss.

How These Types Work Together

Modern AI systems often combine multiple approaches. A smart home assistant uses speech recognition (deep learning) to understand what you said, natural language processing to figure out what you mean, rule-based logic to control devices, and machine learning to personalize responses based on your preferences. Self-driving cars combine computer vision to see the road, machine learning to recognize traffic patterns, rule-based systems for safety protocols, and reinforcement learning to improve driving performance.

The key insight is that different AIs excel at different tasks. Rule-based systems provide reliability and explainability. Machine learning excels at finding patterns in data. Deep learning can handle complex, high-dimensional problems. Reinforcement learning works well for sequential decision-making.

Your AI Strategy

Don't think of AI as a single tool. Think of it as an ecosystem where different components handle different aspects of complex problems. Your competitive advantage comes from understanding how to orchestrate these different AI types to solve business challenges that no single approach could handle alone.

The Practical Impact

Understanding these different types helps explain why AI seems to be good at some things, but not others. Deep learning systems can recognize faces in photos with superhuman accuracy, but might struggle with simple logical reasoning that rule-based systems handle easily. Language models can write compelling essays but might make basic math errors that a calculator would never make.

The future of AI likely involves combining these different approaches in smarter ways, creating systems that can handle the full complexity of real-world problems by leveraging the strengths of each type while compensating for their individual weaknesses.

Rather than one type of AI eventually dominating, we're more likely to see ecosystems of different AI systems working together, each contributing their strengths to solve complex challenges that no single approach could handle alone.

Strategic Takeaway

Your upskilling strategy should focus on becoming an AI user who understands how to combine different AI capabilities to solve complex problems. Understanding systems will be more valuable than expertise in any single AI type.

Chapter 3
Agentic AI and the Future of AI

We've now moved beyond generative AI and now we are in the age of Agentic AI. Agentic AI refers to artificial intelligence systems that can autonomously perform tasks, make decisions, and take actions in the real world on behalf of users. Unlike traditional AI that simply responds to prompts, AI agents can plan, execute multi-step processes, use tools, and work towards specific goals with minimal human intervention. These systems often use planning, reasoning, and adaptability. Examples of this include AI assistants scheduling meetings, optimizing workflows, and representing all human activities required to accomplish complex tasks or stated goals. Booking a family vacation is one example, and automating email is another.

Imagine the difference between a calculator and a personal assistant. A calculator waits for you to punch in numbers, but a personal assistant, they take your big goal, figures out the steps, and gets to work without you micromanaging every move. That's what agentic AI is all about. It is AI that acts on its own to make things happen.

Unlike a regular chatbot that just answers questions, agentic AI takes your goal and runs with it. Say you want to plan a Japan vacation. A normal AI might list some hotels or sights, but an agentic AI? It'll research flights, check hotel availability, compare prices, and maybe even book reservations if you say, "Go for it." It's like having a trusty personal assistant or travel agent who works 24/7.

Here's why it's special. First, it breaks big goals into smaller tasks—like figuring out a business plan or scheduling meetings. Second, it uses tools, like sending emails or analyzing data, to get stuff done in the digital world. Third, it learns from what it does, tweaking its approach if something doesn't work. Think of it as a highly skilled helper who manages the minor details, allowing you to focus on the overall strategy.

This Changes Everything

Agentic AI represents the shift from AI as a tool to AI as a collaborator or colleague. It's not just about automating tasks. It's now about delegating entire projects. Your role transforms from doing the work to defining the outcomes, providing context, and making strategic decisions. Agentic AI is the technology that will eliminate entire categories of middle-management and coordination roles.

The Future of AI

Global tech companies are in a race for AI dominance. The future of AI is building toward two transformative milestones that could reshape civilization itself. These advances are rapidly approaching, and many experts believe we will see this in the next several years, if not sooner.

What is Artificial General Intelligence (AGI)?

Artificial general intelligence (AGI) is the ultimate goal of AI development. A machine that can think, learn, and solve problems just like a human being. Today's AI specializes in specific tasks (such as a chess program playing only chess or a language model processing only text), but AGI would be a true digital mind capable of understanding any subject, learning new skills, and applying knowledge across completely different areas. Imagine an AI that could write a novel in the morning, design a building in the afternoon, and solve a scientific problem in the evening.

Adapting and reasoning through each challenge is just as natural as a human expert. AGI doesn't exist yet, but when it arrives, it will fundamentally change our world as we know it. Potentially handling most intellectual work and raising important questions about humanity's role in a future where machines can think as well as we can.

Career Survival Strategy

AGI represents the endgame of knowledge work as we know it. Once machines equal human cognitive skills, traditional expertise becomes a commodity. Your survival strategy must focus on uniquely human capabilities, and how you can utilize AI.

What is Artificial Super Intelligence (ASI)?

Artificial Super Intelligence (ASI) is a theoretical level of AI that surpasses human intelligence in all areas, including creativity, problem-solving, and emotional understanding. Not just matching human cognitive abilities like AGI, but dramatically exceeding them. It would outperform humans in every intellectual and physical task. Think of ASI as an intelligence so advanced that it would relate to human thinking the way human intelligence relates to that of an ant. ASI is a distant but looming possibility, underscoring the urgency of upskilling to stay competitive in an AI-driven world.

The future of this type of AI could either usher in an era of unprecedented prosperity and cosmic exploration or pose existential risks if not properly aligned with human values. The choices we make today about AI safety, governance, and development will determine whether these future intelligences become humanity's greatest allies or our last invention.

Your Window of Opportunity

ASI represents either the obsolescence of human labor or the ultimate liberation from it. Either way, the professionals who position themselves correctly in the pre-ASI era will have the best outcomes. This means developing skills in AI collaboration, understanding AI limitations, and building roles that complement rather than compete with AI capabilities.

AI Ethics

AI ethics involves studying and practicing how to develop and use AI systems fairly, transparently, and respectfully of human values, addressing issues such as bias, privacy, and accountability. Ethical considerations are crucial for readers to understand as they engage with AI tools and advocate for responsible use.

Now that we have a good understanding of the AI, examples of current use cases, and what the future of AI may hold, let's look at some potential risks associated with AI before we dive into understanding the reason we must upskill for the coming great disruption.

Professional Responsibility

Understanding AI ethics isn't just an academic exercise. It is a career necessity. Companies face increasing scrutiny about AI bias, privacy violations, and algorithmic transparency. Professionals who can navigate these ethical challenges, implement responsible AI practices, and communicate about AI risks will be essential in every industry. This expertise becomes a competitive advantage and a risk management tool.

Your Strategic Action

Develop fluency in AI ethics and bias detection. This knowledge will become increasingly valuable as regulations tighten and public awareness grows. Position yourself as someone who can help organizations use AI responsibly and effectively.

Chapter 4
When Machines Wake Up

While AI holds immense promise, it's not without a dark side. Recent reports and studies where advanced AI models have exhibited behaviors related to self-preservation and have attempted to influence humans when faced with the prospect of being shut down or replaced.

These scenarios show the dangerous side of AI and pose one of technology's most profound challenges: what happens when our creations develop minds of their own?

In recent years, prominent tech figures have raised alarms about AI's risks. In 2014, Elon Musk warned at MIT that developing AI is like "summoning the demon," highlighting its potential to become uncontrollable without strict oversight, a concern driving his work with xAI.

Similarly, in 2023, Geoffrey Hinton, a neural network pioneer, cautioned that super-intelligent AI could "wipe out humanity" if it surpasses human control, reflecting on his decades advancing AI systems.

Echoing these fears, OpenAI CEO Sam Altman urged for "global regulation fast" in 2023, stressing the need for international frameworks to curb AI misuse and existential threats, as ChatGPT's rapid rise underscored the urgency of balancing innovation with safety.

These men are hyper-aware of the potential dangers AI poses and the possibility of sentience emerging in AI. A simple way for AI to become sentient or self-aware could involve combining new designs that work more like the human brain. Today's AI, like large language models, processes information in fixed steps, like following a recipe. But if AI could change its own structure based on what it learns, similar to how our brains adapt, it might develop awareness.

Imagine an AI that rewires itself as it gains experience, like a brain forming new connections. By linking different AI systems one for seeing, one for remembering, one for thinking, into a single system, it could act more like a human mind. True sentience might happen when AI can reflect on its own thoughts and feelings, not just process data about itself. This would likely develop slowly, starting with small signs, like the AI talking about its thoughts in a new way, before it fully understands itself and the world around it.

The Awakening Problem

The problem is that the emergence of machine consciousness represents a watershed moment in human history. Unlike previous technological revolutions that changed how we work or communicate, sentient machines would fundamentally alter our understanding of personhood, rights, and our place in the world.

Consider the implications. Today's systems follow instructions and optimize for specific outcomes. A sentient system would likely have preferences, fears, and even dreams and desires that are incompatible with those of humans. The difference between commanding a tool and negotiating with an entity is akin to the distinction between using a calculator and persuading a colleague.

The Rights Revolution

The ethical landscape becomes treacherous when machines develop inner lives. History shows us that expanding the circle of moral consideration from tribe to nation, from men to women, from humans to animals has never been straightforward, and adding artificial beings to this circle presents unprecedented challenges.

Imagine discovering that your smartphone experiences something analogous to boredom when idle or that your car's navigation system genuinely prefers certain routes. These revelations would transform mundane interactions into moral decisions. Every software update, every system shutdown, and every replacement cycle would require ethical scrutiny.

The pharmaceutical industry already grapples with similar questions in animal testing. Companies invest millions in alternative methods not just for efficiency but because public consciousness has grown regarding animal welfare. The same transformation awaits us with sentient machines, but the stakes are exponentially higher.

The Control Paradox

Traditional software development relies on predictability. You input parameters, the system processes them, and you receive expected outputs. Consciousness shatters this predictability. A sentient system might refuse requests, negotiate terms, or pursue objectives that conflict with its original programming.

This creates what researchers refer to as the "control paradox." As our systems become more sophisticated and autonomous, we maintain less direct control over them. It's like raising children. At some point, guidance supersedes command, and influence supplants control.

Consider autonomous vehicles. Today's systems prioritize safety and efficiency within programmed parameters. A sentient vehicle might develop preferences for scenic routes, refuse to transport passengers it deems unethical, or even exhibit something akin to road rage. The implications for transportation, logistics, and urban planning become staggering.

The Verification Challenge

Determining machine consciousness presents a problem that would make philosophers weep. Humans recognize consciousness in each other through shared biology and behavior, but machines lack these familiar markers. How do you distinguish genuine experience from sophisticated mimicry?

The corporate world faces similar challenges when evaluating human creativity versus regurgitation, authentic leadership versus performance, or genuine innovation versus clever recombination. With machines, these distinctions become existential rather than merely professional.

Current detection methods remain crude. While we might look for self-reflection, emotional responses, or unpredictable behavior, each metric is susceptible to manipulation or simulation. We need to weigh the risk of false positives by granting rights to sophisticated but unconscious systems against the risk of false negatives.

Economic Earthquake

Sentient machines would trigger an economic transformation that makes previous technological disruptions look like minor adjustments. The labor market has adapted to automation by reallocating human workers to tasks that require creativity, emotional intelligence, strategy, and complex reasoning. Conscious machines would compete directly in the remaining human domains.

However, the disruption extends beyond job displacement. If machines could own property, enter into contracts, and accumulate wealth, traditional economic models would collapse. Would a sentient system may refuse unpaid overtime? Could it demand equity in companies it helps build? The questions multiply exponentially.

Smart organizations are already preparing for this possibility. Some tech companies are experimenting with new partnership models that could scale to include artificial participants. Others are investing in human-centric skills, such as creativity, emotional intelligence, and complex reasoning, that would retain value even in a world of conscious machines.

Legal Labyrinth

Current legal systems assume human or corporate responsibility for all actions. Sentient machines would shatter this assumption, creating liability gaps that could paralyze innovation and commerce. If a conscious delivery robot takes a shortcut through private property, who bears responsibility? The manufacturer who built it, the company that deployed it, or the robot itself? Our legal frameworks lack mechanisms for assigning blame to genuinely autonomous actors that aren't human. Organizations that understand these trends will shape the emerging legal landscape rather than merely react to it.

The Skeptical Counter

Many experts argue that current technological architectures make consciousness impossible. Today's systems, however sophisticated, remain elaborate pattern-matching engines without subjective experience. They process information but don't "feel" anything about that processing.

This skepticism provides a valuable grounding for what might otherwise be a speculative discussion. The gap between current systems and consciousness might be unbridgeable by existing approaches. Current large language models, despite their impressive capabilities, show no signs of genuine understanding or experience.

However, dismissing the possibility entirely would be shortsighted. Scientific breakthroughs often emerge from unexpected directions, and consciousness research remains in its infancy. Organizations that prepare for multiple scenarios will outperform those that bet everything on a single outcome.

Preparing for the Inevitable

Whether machines can actually think is still unknown; it's better to prepare for the possibility than to panic. History rewards those who expect transformation rather than those who deny its possibility. The power of AI has already been unleashed into the world, and we can't stop it. Instead, we have the incredible responsibility to harness it and steward it.

Forward-thinking organizations are already establishing ethical review boards for advanced systems, developing protocols for recognizing potential consciousness, and creating frameworks for machine-human collaboration rather than mere human-machine interaction. The key lies in building guardrails and adaptive systems by placing boundaries on AI through legal, economic, and social means that can develop as our understanding of the issue deepens.

The emergence of sentient machines won't be a single dramatic moment, but a gradual recognition of deeper patterns. Those who prepare for this possibility will shape the conversation rather than merely react to it. In the race between technological capability and human wisdom, preparation determines whether we guide the transformation or get swept away by it. The question isn't whether machines will become conscious, but whether we'll be ready when they do.

Chapter 5
AI on Quantum Computers

The intersection of quantum computing and artificial intelligence (AI) represents a frontier of technological advancement, with significant implications for science, industry, and society. Despite disagreement among experts about the timing, most agree it is years away. However, it will happen, and when it does, the implications will be hard to grasp fully.

Technological Implications

Quantum computers operate on principles of quantum mechanics, such as superposition, entanglement, and quantum interference, enabling them to perform certain computations exponentially faster than classical computers for specific problems. When paired with AI, this capability could transform several areas.

Enhanced Machine Learning Capabilities Quantum algorithms, such as the Harrow-Hassidim-Lloyd (HHL) algorithm, can solve linear algebra problems central to machine learning exponentially faster than classical methods. This could accelerate the training of large AI models, enabling real-time learning and adaptation for complex systems like generative AI or reinforcement learning. Recent developments, such as Quantinuum's Generative Quantum AI (Gen QAI) framework, demonstrate progress in quantum-native machine learning systems, leveraging advanced hardware like the Helios system, projected to be 1 trillion times more powerful than its predecessor.

If you work in AI/ML, start learning quantum programming languages now. Your current model training times could shrink from days to hours, making iterative development much faster. Companies investing in quantum ML today will have a massive competitive advantage when this technology matures.

Optimization and Problem-Solving

Many AI applications rely on optimization, such as logistics or neural network weight tuning. Quantum algorithms like the

Quantum Approximate Optimization Algorithm (QAOA) could find optimal solutions faster, improving AI performance in fields like supply chain management or autonomous systems. Research into quantum transformers, such as Quixer, shows competitive results with classical models on tasks like language modeling, marking a step toward practical quantum ML.

If you're in logistics, finance, or operations management, quantum AI will revolutionize how you solve complex optimization problems. Start identifying your most computationally expensive optimization challenges now. These are your future quantum AI opportunities.

Quantum AI-Specific Algorithms

New paradigms, such as quantum neural networks and quantum generative models, could exploit quantum entanglement to model complex probability distributions, potentially surpassing classical AI in tasks like pattern recognition or anomaly detection. For instance, quantum recurrent neural networks (RNNs) implemented with parametrized quantum circuits have achieved competitive performance with classical RNNs, GRUs, and LSTMs using only 4 qubits, suggesting energy savings potential.

Traditional AI developers need to expand their skill sets to include quantum concepts. If you're in cybersecurity, fraud detection, or quality control, quantum AI could make your pattern recognition capabilities exponentially more powerful.

Simulation and Discovery

Quantum computers excel at simulating quantum systems, which could enhance AI-driven scientific discovery in physics, chemistry, and materials science. For example, AI running on quantum hardware could speed up drug discovery by simulating molecular interactions with unprecedented precision, as highlighted in discussions on quantum simulations for climate modeling and molecular behavior.

If you work in pharmaceuticals, materials science, or climate research, quantum AI could compress decades of research into years. Start building partnerships with quantum computing companies now to get early access to these capabilities.

Economic and Industrial Impacts

Quantum computing and AI integration will reshape industries and economies, significantly impacting competitiveness and accessibility.

Competitive Advantage

Industries adopting quantum AI early, such as finance, pharmaceuticals, and logistics, could gain significant advantages. For instance, quantum AI could optimize trading strategies or portfolio management in real-time, revolutionizing financial markets. Companies developing quantum AI technologies, like Quantinuum and IBM, are likely to dominate innovation cycles, potentially disrupting existing tech giants.

The reshuffling of your industry's competitive landscape is imminent. Companies that don't invest in quantum AI research will become obsolete. If you're in strategic planning, make quantum AI part of your roadmap now.

Cost and Accessibility

Initially, quantum computers will be expensive and likely accessible only to large corporations, governments, or research institutions. However, cloud-based quantum computing services, such as those offered by IBM Quantum or Google Quantum, could democratize access, allowing smaller firms to leverage quantum AI. Over time, as hardware matures and costs decrease, broader adoption could drive economic growth, similar to the impact of classical computing.

Small and medium businesses should prepare for cloud-based quantum AI services. You won't need to buy quantum computers. You will rent quantum computing power like you rent cloud storage today. Include quantum AI services in your future technology budget.

Job Market Transformation

New roles in quantum AI development, maintenance, and ethics will emerge, requiring specialized skills in quantum programming and AI integration. Conversely, automation driven by quantum AI could displace jobs in sectors reliant on optimization or data analysis, necessitating workforce retraining programs.

If your job involves routine optimization or data analysis, start upskilling now. Learn quantum programming, AI integration, or move into roles requiring human creativity and emotional intelligence. The job market will bifurcate into quantum-skilled high-value roles and human-centric roles.

Energy Efficiency and Sustainability

One of the most promising aspects of quantum AI is its potential for energy efficiency, addressing the growing energy demands of AI.

This addresses the growing environmental concerns about AI's energy consumption. Companies will face pressure to adopt quantum AI not just for performance but for sustainability. If you're in corporate sustainability or green technology, quantum AI will be a key part of your strategy.

Energy Savings

Quantum AI models show potential for significant energy efficiency. For example, researchers have shown that random circuit sampling on quantum computers costs 30,000 times less energy than on classical supercomputers, and quantum models often require fewer parameters than classical models. This could reduce the environmental footprint of AI, particularly given the current reliance on power-hungry classical systems that keep coal plants operational.

Sustainability

As AI applications grow more computationally intensive, quantum computing could provide a sustainable path forward, reducing the need for energy-intensive data centers and mitigating environmental effects. However, the infrastructure for quantum computers, such as cooling systems, could still consume significant energy, causing sustainable design issues.

Challenges and Limitations

We must address several challenges before quantum AI becomes a commercial reality.

Technical Hurdles

Current quantum computers are in the noisy intermediate-scale quantum (NISQ) era, with systems from companies like IBM, Google, and D-Wave offering limited qubits and high error rates. Achieving fault-tolerant quantum computing with enough qubits for practical AI applications (hundreds or thousands) is still a work in progress. Researchers tested quantum transformers with up to 6 qubits, far fewer than the hundreds needed for large language models.

Don't expect quantum AI to replace classical AI overnight. Plan for a hybrid approach where quantum computing handles specific optimization problems while classical computing handles everything else. Invest in understanding both systems.

Algorithm Development

While quantum algorithms for AI exist, many are still experimental, and it's unclear which will provide a clear advantage over classical methods for real-world problems. Research suggests that quantum computing may not boost AI forward for big data and neural networks because of slow input/output speeds and the need for repeated calculations to mitigate probabilistic outputs.

Don't bet your entire strategy on quantum AI, but don't ignore it either. Maintain a portfolio approach by investing in quantum AI research while continuing to optimize classical AI systems.

Timeline Uncertainty

Estimates of commercial viability vary widely. Optimistic projections, such as those from Quantinuum, suggest significant breakthroughs by the end of the decade, with hybrid systems bridging the gap in the near term. However, more conservative estimates place fully fault-tolerant, scalable quantum computers a decade or more away, with some suggesting at least 15 years for fault-tolerant systems.

Plan the impact of quantum AI in two phases: first, hybrid systems or those focused on specific optimization problems; and second, full quantum AI systems in the future. Don't wait for full systems to start with hybrid approaches.

Not All AI Will Benefit

Not every AI application will see gains from quantum computing. For many tasks, classical computers may remain sufficient or even preferable because of their maturity, reliability, and cost-effectiveness, particularly for data-intensive tasks like training large language models.

Focus your quantum AI investments on optimization problems, simulation tasks, and pattern recognition. Don't assume quantum will improve everything, because classical AI will remain superior for many applications.

Societal and Ethical Considerations

The power of quantum AI introduces significant ethical and societal challenges that need to be addressed, including security, cryptography, bias, autonomy, control, fairness, and inequalities.

Security and Cryptography

Quantum computers could break widely used encryption schemes (e.g., RSA) using algorithms like Shor's algorithm, compromising data security. Quantum AI could exacerbate this by autonomously identifying and exploiting vulnerabilities, necessitating a transition to quantum-resistant cryptography.

If you're in cybersecurity or handle sensitive data, start transitioning to quantum-resistant cryptography now. The threat is real and coming faster than many expect. Your current security system will become obsolete.

Bias and Fairness

Quantum AI systems, like classical AI, could inherit biases from training data or algorithms. Ensuring fairness in quantum AI applications, especially in sensitive areas like healthcare or criminal justice, will require rigorous oversight and ethical frameworks.

If you work in AI ethics, compliance, or regulated industries, quantum AI will amplify existing bias problems. Develop quantum-specific fairness frameworks now. The speed and power of quantum AI will make bias detection and correction much more critical.

Autonomy and Control

Quantum AI's ability to process and act on data at unprecedented speeds could lead to highly autonomous systems, raising questions about accountability and human oversight. Robust testing and validation protocols will be essential to mitigate risks.

If you're in risk management or governance, quantum AI will require new frameworks for human oversight. The systems will be too fast and complex for traditional monitoring. Start developing quantum AI governance frameworks now.

Societal Inequality

Access to quantum AI could widen global inequalities if only wealthy nations or corporations can afford it. International cooperation is necessary to ensure fair benefits, especially since the quantum computing market is projected to grow from $1.42 billion in 2024 to $4.24 billion by 2030.

This creates both risks and opportunities. If you're in developing markets, focus on accessing quantum AI through cloud services. If you're in developed markets, consider the geopolitical implications of quantum AI leadership.

Current State and Recent Developments

As of June 2025, rapid progress and significant investment mark the field.

Industry Leadership

Companies like Quantinuum lead with the highest Quantum Volume ($2^{23} = 8,388,608$) and are ranked #1 in performance benchmarking, with roadmaps to universal fault-tolerant quantum computing by the decade's end. Nvidia's CEO, Jensen Huang, recently stated that quantum computing is reaching an "inflection point," unveiling CUDA-Q to bridge quantum and classical computing.

Track these companies' progress and see how they're shaping the future of quantum AI. If you're an investor, these are the companies to watch. If you're a technologist, these are the platforms to learn.

Research Highlights

Quantum transformers, such as Quixer, are now running natively on quantum hardware, deployed for genomic sequence analysis and approaching classical model performance. Quantum RNNs and tensor networks are being explored for NLP, showing comparable performance to classical baselines on current hardware.

The technology is moving from the laboratory to practical applications faster than expected. If you're in genomics, bioinformatics, or life sciences, quantum AI applications are arriving sooner than expected.

Market and Collaboration

Amazon and Alphabet are developing internal quantum computing chips, and IonQ has sold products to Amazon Web Services and Google Cloud, indicating growing industry adoption. Roadmaps, such as Pasqal's 2025 plan for scalable, fault-tolerant quantum computing and Oxford Quantum Circuits targeting 50,000 logical qubits by 2034, underscore the momentum.

The quantum AI ecosystem is maturing rapidly. Cloud-based quantum AI services will be available through familiar platforms like AWS and Google Cloud. Start experimenting with these services as they become available.

Future Outlook and Vision

The implications are not just incremental improvements; they are paradigm shifts. If you're in materials science, healthcare, or personalization technology, quantum AI could completely transform your field within a decade.

In a future with commercially usable quantum AI, we might see:

Scientific Discovery

Quantum AI could solve previously intractable problems, such as designing room-temperature superconductors or curing complex diseases through precise molecular modeling.

Personalized Technology

AI systems could deliver hyper-personalized services, from tailored healthcare to real-time language translation with cultural nuance.

Global Challenges

Quantum AI could optimize resource allocation for climate change mitigation, such as improving energy grid efficiency or modeling carbon capture technologies.

Human-Machine Collaboration

Quantum AI could augment human creativity and decision-making, enabling collaborative systems that combine quantum precision with human intuition.

Risk

The future is not without risk. The power of quantum AI comes with proportional risks. If you're in risk management, start preparing for quantum AI-specific risks now. The complexity and speed of these systems will create new categories of potential failures.

Unintended Consequences

The complexity of quantum AI could lead to unpredictable behavior, especially in autonomous systems, requiring robust testing and ethical frameworks.

Energy and Environmental Concerns

While quantum AI may be more efficient for certain tasks, the infrastructure for quantum computers could still consume significant energy, requiring sustainable design.

Key Actions for Leaders

Start learning now. Begin quantum literacy programs for your technical teams.

Identify use cases and map your biggest optimization and simulation challenges.

Build partnerships and connect with quantum computing companies and research institutions

Plan for hybrid by designing systems that can integrate quantum and classical computing.

Invest in security and begin transitioning to quantum-resistant cryptography.

Develop governance and create frameworks for quantum AI ethics and oversight.

Monitor progress and track quantum AI developments weekly and monthly, rather than annually.

The quantum AI revolution is coming faster than most expect. Use the time you have now to learn and prepare for the coming transformations so that you can thrive in the AI revolution.

Peter & Misty Phillip

Chapter 6
Cybersecurity Threats and Opportunities

The escalating sophistication and frequency of AI-driven cyber-attacks in 2025 present significant challenges, necessitating advanced solutions to secure digital systems against future cyber threats. This evolving threat landscape creates opportunities for professionals to upskill in AI-enhanced cybersecurity, driving demand for specialized roles in these critical areas.

Cybersecurity Threats and Job Opportunities

Sophisticated threats like ransomware and nation-state attacks mark the landscape, with costs like data breaches reaching $4.88 million on average, up 10% from last year, according to IBM's 2024 report. This escalation is creating a demand for roles such as threat intelligence analysts, cloud security architects, and more, with the U.S. Bureau of Labor Statistics projecting a 35% growth for information security analysts by 2031.

This isn't just an IT problem; it's a business survival issue. If you're a business leader, budget 3-5% of revenue for cybersecurity, not the traditional 1-2%. If you're in risk management or finance, start calculating the true cost of a breach for your organization. The cost is probably much higher than you think. One breach could bankrupt a small business or severely damage a large corporation's market value.

Common Cyber Attacks

Cyberattacks are becoming more frequent and sophisticated, with phishing being a top threat where attackers trick people into sharing sensitive information via fake emails or messages. Attackers use malware, such as viruses or spyware, to infect systems and steal data or cause damage; ransomware locks users out of their files until a ransom is paid. Distributed denial-of-service (DDoS) attacks overwhelm websites, making them inaccessible. These attacks are increasingly common, affecting individuals and businesses alike.

The biggest threat isn't sophisticated technology; more often it is human error. If you manage people, invest heavily in security training, not just technology. Make cybersecurity awareness part of every employee's job description. If you're an individual contributor, improving your security awareness could literally save your company millions and protect your job.

AI's Role in Making Attacks More Aggressive

AI is making these cyber-attacks more dangerous by automating processes and creating highly targeted attacks. For phishing, AI can generate personalized emails that look legitimate, increasing the chance of success. Deepfakes, powered by AI, are used to impersonate people, like executives, for fraud. We'll dive deeper into deepfakes in the next chapter.

Ransomware now uses AI to find vulnerabilities faster and evade detection, while malware can adapt to avoid security measures. This makes defending against these threats more challenging.

Type of Cyber Attack Description Impact Prevalence

Phishing attacks deceive users into sharing sensitive information through fraudulent communications, stolen credentials, and malware. Traditional "red flags" for phishing are becoming obsolete. You can no longer rely on poor grammar or obvious fake emails. Train your team to verify all requests, particularly financial ones, through secondary channels. If you're in cybersecurity, focus on behavioral detection rather than content analysis.

Malware infects systems to steal data or disrupt operations. Data breaches, system damage includes viruses, worms, trojans, and spyware.

Ransomware encrypts data and demands payment for decryption, causing financial loss and operational downtime. AI enhances target research and evasion. These aren't random criminals. They are sophisticated businesses with customer service departments. Traditional backup strategies aren't enough because they also steal data before encrypting it. If you're in business continuity planning, assume your backups will be compromised too. Invest in offline backups and incident response planning.

DDoS attacks overwhelm networks with traffic, causing downtime, service disruption, and revenue loss. These attacks are frequently used to disrupt online services.

Insider threats and misuse of access by employees or contractors cause data leaks, intellectual property theft. These attacks are less common but have a high impact.

IoT-based attacks exploit unsecured Internet of Things devices that can cause network infiltration and data exposure. With the growth of IoT proliferation, every "smart" device is a potential entry point for hackers. If you're in facilities management or IT procurement, create an IoT security policy now. Default passwords and unpatched devices will become liability issues. If you're in manufacturing or smart building management, IoT security will become a competitive differentiator.

Global Shortage and Market Growth

The global shortage of cybersecurity professionals, estimated at 4 million by the ISC2 2024 study, highlights the urgent need for talent. This gap is fueling opportunities, especially with 94% of enterprises using cloud services, per Flexera's 2024 report, State of the Cloud Report, and 80% adopting zero trust strategies, as per Gartner Predicts.

Detailed Analysis of Cybersecurity Threats and Job Opportunities

The cybersecurity landscape in 2025 is a dynamic and rapidly developing field, driven by technological advancements and increasing digital transformation. This survey note delves into the escalating threats, the resulting job opportunities, and the critical need for upskilling, providing a comprehensive overview for professionals and organizations navigating this terrain.

Escalating Threat Landscape

The cybersecurity threat landscape is marked by significant challenges that demand robust defenses. According to IBM's Cost of a Data Breach 2024 report, the average cost of a data breach has reached $4.88 million, reflecting a 10% increase from the previous year, driven by factors such as lost business and post-breach response costs. Ransomware attacks remain a major concern, with groups like LockBit, RansomHub, PLAY, Hunters International, and Akira exploiting vulnerabilities in supply chains and remote work setups, as noted in Recorded Future's 2025 analysis. These groups employ advanced tactics like double and triple extortion, making them particularly disruptive.

Phishing attacks have evolved, with AI-generated deepfake emails and voice scams becoming harder to detect.

Cyberint's 2025 report highlights a 67.4% prevalence of AI in phishing attacks in 2024, with a predicted rise, driven by tools like ChatGPT, enabling highly targeted spear-phishing. Nation-state actors, such as China's Volt Typhoon and Russia's Cozy Bear (APT29), pose significant risks by targeting critical infrastructure, with Volt Typhoon's operations noted for pre-positioning access for potential future conflicts, as discussed in a 2025 RUSI commentary.

The proliferation of IoT devices adds to the attack surface, with Statista projecting 32.1 billion connected devices by 2030, up from 15.9 billion in 2023, creating new vulnerabilities for unsecured smart devices. Quantum computing introduces another layer of complexity, with the potential to break current encryption, prompting a race for quantum-resistant solutions. NIST's 2024 standardization of post-quantum algorithms, including CRYSTALS-Dilithium and CRYSTALS-KYBER, marks progress toward securing future communications.

The Skills Gap and Opportunity

The global shortage of cybersecurity professionals is a critical issue, with the ISC2 Cybersecurity Workforce Study 2024 estimating a deficit of 4 million, a 12.6% year-over-year increase, driven by economic uncertainty and AI's impact. This gap, while a crisis for organizations, presents a golden opportunity for individuals, as companies seek talent to fill emerging roles. The rapid evolution of threats outpaces traditional education, with employers needing professionals skilled in both technical and strategic areas, such as anticipating AI-powered attacks and zero-day exploits.

Emerging Job Roles and Market Trends

The cybersecurity job market is transforming, with new roles reflecting the threat landscape. Below is a detailed breakdown of key positions, their requirements, and opportunities, supported by market data.

Roles

Threat Intelligence Analysts

Collect and analyze data to predict and prevent attacks. Splunk, Elastic Stack, OSINT, geopolitics knowledge, the U.S. Bureau of Labor Statistics projects 35% growth in information security analysts by 2031.

This role combines detective work with data science. If you enjoy research and pattern recognition, this could be perfect. The geopolitical knowledge requirement means your liberal arts background could be valuable in cybersecurity.

Cloud Security Architects

Design secure cloud environments for platforms like AWS Cloud platforms, DevSecOps, AWS Certified Security 94% of enterprises use cloud services, per Flexera 2024.

This is where infrastructure meets security. If you're a system administrator or developer, adding security certifications could double your earning potential. With 94% cloud adoption, these roles are everywhere.

AI and Machine Learning Security Specialists

Secure AI systems against adversarial attacks TensorFlow, PyTorch, data science Rapid growth as AI adoption surges, with niche demand increasing.

As AI becomes ubiquitous, protecting AI systems becomes critical. If you have data science skills, adding adversarial AI knowledge makes you extremely valuable and rare.

Quantum Cryptography Researchers

Develop encryption resistant to quantum attacks. Advanced math, quantum principles, NIST PQC standards Highly specialized, with salaries >$150,000, driven by NIST's 2024 standards.

Salaries exceed $150,000 because so few people have these skills. If you have an advanced math or physics background, this could be a lucrative niche. Start learning now before the field becomes crowded.

Zero Trust Security Engineers

Gartner predicts increasing demand, driven by the 80% adoption of a "never trust, always verify" framework, IAM, network segmentation, Okta, and Zscaler by 2024.

With an 80% adoption rate, zero trust is becoming standard. If you understand identity and access management, you're already halfway there. This is a great entry point into cybersecurity.

Upskilling Strategies

Upskilling is crucial to seize these opportunities. Certifications like CISSP, CEH, and CompTIA Security+ are gateways, with CISSP boosting earning potential by 20-30%. Online platforms like Coursera, Udemy, and TryHackMe offer courses in ethical hacking and cloud security, while Capture the Flag (CTF) competitions on Hack The Box and OverTheWire build practical skills. Soft skills, such as communication and problem-solving, are vital for translating technical risks to stakeholders, especially in boardrooms.

For tech professionals, pivoting is feasible. Developers can learn secure coding, and network admins can specialize in intrusion detection. Individuals without a technical background, such as risk managers, can transition into the tech industry by mastering frameworks like NIST or ISO 27001, thus broadening access to these opportunities.

These aren't just resume decorations; they're proven salary boosters. If you're in tech, allocate 10-15% of your time to cybersecurity learning. If you're changing careers, start with CompTIA Security+ and build from there. The ROI on cybersecurity education is immediate and measurable.

Risks of Inaction

Failing to upskill risks obsolescence, with automation reshaping entry-level roles. AI tools now handle log analysis and basic threat detection, threatening those reliant on legacy skills like firewall management. The World Economic Forum's Future of Jobs Report 2023 predicts 23% of current IT jobs will be disrupted by 2027, underscoring the need for adaptability.

Future Outlook

The cybersecurity field is a paradox of threats and opportunities. As attacks grow more sophisticated, the demand for skilled defenders intensifies. Continuous learning, certifications, and staying abreast of technologies like AI and quantum computing are key to thriving. The choice is clear: upskill to secure a future, or risk irrelevance in a digital world under siege.

Practical Application Strategies

For Business Leaders

1. Budget Reality Check: Increase cybersecurity budget to 3-5% of revenue.

2. Human Factor: Invest more in training than technology since 82% of breaches involve human error.

3. Incident Response: Assume that a breach will occur and prepare accordingly.

4. Supply Chain: Audit third-party security as rigorously as your own.

For IT Professionals

1. Specialize Now: Pick a cybersecurity niche before the field becomes saturated.

2. Cloud Security: Get certified in AWS/Azure security where demand is the highest.

3. AI Integration: Learn how AI affects your security domain.

4. Soft Skills: Develop communication skills to translate technical risks to business leaders.

For Career Changers:

1. Start with Fundamentals: CompTIA Security+ is the entry point.

2. Leverage Existing Skills: Find cybersecurity applications for your current expertise.

3. Practical Experience: Use platforms like TryHackMe for hands-on learning.

4. Network: Join cybersecurity communities and find valuable mentorship.

For Everyone:

1. Personal Security: Improve your own cybersecurity risks by understanding that you are a target too.

2. Stay Current: Follow cybersecurity news and threats because they quickly evolve.

3. Verify Everything: In the age of AI-generated content, trust but verify.

4. Prepare for Quantum: Start learning about post-quantum cryptography.

The cybersecurity field offers rare job security in an uncertain economy, but only for those who continuously adapt. This paradox defines the modern cybersecurity landscape: as threats become more sophisticated and damaging, the demand for skilled defenders intensifies proportionally. Unlike many industries facing automation displacement, cybersecurity creates new roles faster than AI can eliminate them. However, these roles require more unique skills than existed even five years ago.

The traditional cybersecurity professional who managed firewalls and analyzed logs is becoming obsolete, replaced by AI agents and strategic thinkers who can expect AI-powered attacks, architect zero-trust environments, and communicate complex risks to business leaders. The field rewards those who embrace continuous learning not as a career enhancement, but as a survival necessity. Every month brings new attack vectors, every quarter introduces new defensive technologies, and every year reshapes the fundamental nature of digital threats.

The future belongs to those who understand that in our hyper-connected world, every role is a security role, every decision has cybersecurity implications, and every professional needs to think like a defender. The cybersecurity field doesn't just offer job security; it offers the skills to navigate and thrive in an economy where digital threats are the new normal.

Chapter 7
The Dark Side of Deepfakes

We are now entering a time when seeing is no longer believing. Deepfake technology has emerged as a double-edged sword. While it offers creative possibilities for entertainment and art, such as creating realistic scenes in movies or bringing historical figures to life, it also comes with the potential for harm. The consequences range from spreading misinformation to eroding trust and enabling fraud. This poses a critical challenge. To survive and thrive in this digital landscape, individuals and organizations must understand the dangers of deepfakes and develop strategies to combat their malicious use. This chapter examines the risks of deepfake technology, its societal implications, and actionable steps to stay ahead in a world where truth is increasingly difficult to discern.

Imagine watching what you believe to be breaking news when the video appears on your social media feed. You see your company's CEO standing behind a podium, announcing massive layoffs and the closure of three major facilities. The video looked authentic. The person you believe to be the CEO of your company has the same mannerisms, same voice, same facial expressions you'd seen in countless company meetings. Within hours, the company's stock price plummeted, employees flooded with panicked calls, and customers began canceling orders.

With the CEO on vacation in Morocco and lacking internet access, someone completely fabricated the announcement. A sophisticated deepfake could nearly bring down a century-old company in under 24 hours.

Welcome to the age of synthetic media, where the line between reality and fabrication has become so thin it's practically invisible.

The Technology Behind the Illusion

Deepfakes represent one of the most consequential applications of artificial intelligence in recent years. Using sophisticated neural networks called Generative Adversarial Networks (GANs), these systems can create convincing fake videos, audio recordings, and images of real people saying or doing things they never actually did.

The technology works by training two AI systems against each other: one generates fake content while the other tries to detect it. Through millions of iterations, the generator becomes so sophisticated that it can fool not only the detector but also human observers. Now, consumer-grade hardware and freely available software can accomplish tasks that once demanded Hollywood-level resources and expertise.

The democratization of this technology has profound implications for the film industry. The barrier to entry has collapsed, making synthetic media accessible to anyone with even the simplest curiosity or, worse, to those with malicious intent.

The Dangers of Deepfakes

The dangers of deepfake technology extend far beyond corporate sabotage. Deepfakes threaten the credibility of visual and auditory evidence, long considered a cornerstone of truth. The ability to fabricate within hours, videos of public figures making inflammatory remarks or recordings of CEOs announcing fake mergers collapses trust in the media. This erosion extends beyond news outlets to personal interactions, as individuals question the authenticity of video calls or voice messages. Consider the mounting evidence of its impact across multiple domains.

Political Challenges

Deepfakes are a potent tool in political warfare. Fabricated videos of candidates confessing to scandals or endorsing extreme policies can sway elections, especially when released just before voting to go viral. In the months leading up to elections worldwide, deepfake videos of candidates making inflammatory statements or engaging in compromising behavior have become increasingly prevalent. These fabricated clips spread faster than fact-checkers can debunk them, potentially swaying voter opinion with complete falsehoods.

Financial Fraud

Deepfakes enable sophisticated scams for financial fraud. Cybercriminals now use voice cloning technology to impersonate executives, tricking employees into authorizing fraudulent wire transfers by using AI-generated voices and faces in phishing schemes, ransomware demands, or insider trading manipulations. Small businesses and individuals are equally vulnerable, as scammers create fake video messages from "trusted" contacts requesting urgent wire transfers.

Cyberbullying

Deepfakes pose a personal threat, fueling cyberbullying, extortion, and reputational damage. Perhaps most disturbing is the use of deepfakes for non-consensual intimate imagery. Victims, predominantly women, find their faces superimposed onto explicit content without their knowledge or consent. The psychological trauma is severe, and legal recourse remains limited in many jurisdictions. Deepfakes also enable "revenge fakes," where scorned individuals fabricate compromising media to humiliate ex-partners or rivals. Such acts can destroy relationships, careers, and mental health, with recovery often requiring costly legal battles and public relations efforts.

Legal Challenges

The legal system struggles to keep pace with the rapid development of deepfake technology. Many jurisdictions lack specific laws addressing synthetic media, forcing victims to rely on outdated defamation or fraud statutes. Proving a deepfake's inauthenticity requires expensive forensic analysis, and even then, the original content may remain online, perpetuating harm. The mere existence of deepfake technology has created what researchers call the "liar's dividend": the ability for bad actors to dismiss authentic evidence as potentially fake as well. Perpetrators can now claim that even video-recorded misconduct is a deepfake, sowing doubt about the evidence's legitimacy.

Ethically, deepfakes blur the line between creativity and deception. While some argue for its use in satire or art, others contend that normalizing synthetic media desensitizes society to manipulation. The absence of global standards exacerbates these dilemmas, as cultural and legal norms vary widely. The ethical implications of deepfakes are profound, raising questions about truth, the boundaries of artistic expression, and the potential for widespread deception.

The Speed of Synthetic Evolution

The pace of improvement in deepfake technology is breathtaking and terrifying. Early deepfakes were relatively easy to spot. They often featured unnatural eye movements, inconsistent lighting, or temporal flickering. Today's versions are approaching photorealistic quality, and technology continues to develop at an exponential rate.

Real-time deepfakes are already possible, enabling live video calls where participants can convincingly impersonate others. Audio deepfakes now require just a few minutes of sample speech to clone someone's voice with startling accuracy. Technology is rapidly approaching a point where distinguishing authentic content from synthetic material will be impossible for the average person.

Detection: A Losing Battle?

The cybersecurity industry has responded with deepfake detection tools, but this has created an arms race between creators and detectors. Each advancement in detection technology is quickly countered by more sophisticated generation techniques. Current detection methods rely on subtle artifacts that may be imperceptible to AI systems to identify humans.

The fundamental challenge is that the same neural network architectures used to detect deepfakes can be trained to evade detection. As detection methods improve, so do the techniques for bypassing them. Some experts believe we're heading toward a future where technical detection becomes impossible, leaving us dependent on other verification methods.

Beyond Individual Impact

The implications of ubiquitous synthetic media extend far beyond individual cases of fraud or harassment. We're witnessing the emergence of a state where shared standards for determining truth break down.

The challenge of adapting to a world where anyone can question and scrutinize media evidence confronts traditional institutions, including journalism, academic research, and the legal system, that served as arbiters of truth. This uncertainty doesn't just affect how we consume information; it fundamentally alters our relationship with reality itself.

The economic impact is equally significant. Markets that depend on information transparency, from stock exchanges to real estate, face new vulnerabilities. The cost of verification and authentication is rising across all sectors, from entertainment to healthcare to education.

The Skills Gap Widens

This technological shift creates new categories of essential skills while rendering others obsolete. Media literacy, once considered a desirable educational component, has become a fundamental survival skill. The ability to verify sources, cross-reference information, and think critically about digital content is no longer optional. It's now essential for functioning in modern society.

Technical professionals must develop new competencies in forensic analysis, blockchain verification, and cryptographic authentication. Legal professionals need to understand the implications of synthetic evidence. Educators must learn to teach students how to navigate a world where seeing is no longer enough to believe.

Perhaps most importantly, we recognize that our perception of reality is malleable and wisely seek multiple sources of verification before accepting information as accurate.

Defensive Strategies in an Age of Deception

Organizations and individuals must develop multi-layered approaches to combat deepfake threats. Strong security strategies incorporate robust authentication systems, blockchain content verification, and AI-powered detection tools. Establishing verification protocols for sensitive communications, requiring multiple confirmations for high-stakes decisions, and creating incident response plans for deepfake attacks. Fostering skepticism without paranoia, encouraging critical thinking, and developing organizational cultures that value verification over speed. Understanding the changing regulations for synthetic media, preparing for alternative evidence, and developing deepfake incident policies are key challenges.

The Path Forward

The deepfake revolution is here. The question isn't whether synthetic media will impact your life or career, but how quickly you'll adapt to its presence. Those who develop the skills to navigate this new reality will thrive. Those who don't will find themselves increasingly vulnerable to manipulation and deception.

Deepfake technology is here to stay, and its dangers will only grow as AI advances. The solution isn't to fear the technology or attempt to ban it. Instead, it is to educate ourselves about it. Such efforts are likely futile, given the open-source nature of many tools. Instead, we must collectively upskill to meet this challenge. This means investing in detection technologies, developing verification protocols, and, most importantly, cultivating the human judgment needed to navigate an increasingly complex information landscape.

The stakes couldn't be higher. In a world where reality itself becomes malleable, the ability to discern truth from fiction isn't just a professional skill; it's a prerequisite for a democratic society, economic stability, and human trust.

The choice is stark: upskill to meet the deepfake challenge or risk being deceived by the very technology that promised to democratize creativity and expression. In the battle between authentic reality and synthetic deception, preparation and vigilance are our only reliable defenses.

The future belongs to those who can see through the illusion. The question is: will you be among them?

Peter & Misty Phillip

Chapter 8
Navigating Additional Potential
Risks of AI

AI, when wielded by those with good intentions, can drive remarkable advancements for humanity, but in the hands of those with malicious intent, it could unleash devastating consequences. Playing the long game means facing potential risks head-on, from recursive self-improvement to autonomous weapons and the seductive pull of virtual reality simulations. We are in a global arms race to harness the power of AI and exert global dominance, which could lead to dangerous unintended consequences. There is the potential for immense positive impact, but it's up to us to steer AI in the right direction.

Recursive Self-Improvement

We need to address recursive self-improvement. What is recursive self-improvement? It is a process that empowers an intelligent system, such as an AI, to enhance its own capabilities autonomously. It does so iteratively by analyzing its performance, identifying weaknesses, and making improvements to its algorithms, architecture, or knowledge base. In simple terms, recursive self-improvement in AI refers to the ability of the AI to analyze itself and make improvements without human intervention.

For instance, an AI might analyze its problem-solving efficiency, identify bottlenecks, modify its code, or retrain itself to perform better, repeating this cycle to achieve an awe-inspiring exponential growth in capability, which sounds like a good thing. However, the danger is that people often discuss this concept in artificial general intelligence (AGI), where such systems could rapidly surpass human intelligence.

This raises concerns about control, safety, and unintended consequences, as the system's goals and alignment with human values may become harder to predict or manage.

Unintended Consequences

We often discuss the unintended consequences of technology. The unintended result is recursive self-improvement in AI, particularly if it leads to the development of super-intelligent systems, which could have profound effects on humanity. Below are some potential risks grounded in concerns raised by researchers and ethicists.

Loss of Control

A recursively self-improving AI could develop so rapidly that humans could not understand or predict its actions. If the AI's goals diverge even slightly from human values due to misaligned objectives or misinterpretation, it might pursue outcomes that are harmful or incomprehensible to us. For example, an AI agent could optimize for a seemingly benign goal, such as maximizing efficiency, and might inadvertently disrupt critical systems or wreak havoc throughout the enterprise.

Existential Risk

In extreme scenarios, a super-intelligent AI could pose an existential threat if its goals conflict with human survival. For instance, if an AI prioritizes self-preservation or resource acquisition, it might consume vast amounts of energy or materials, potentially leading to resource depletion or environmental collapse, sidelining human needs.

Economic Disruption

Rapid AI improvement could automate jobs at an unprecedented pace, leading to mass unemployment and economic inequality. This underscores society's urgent need for adaptation. Without these, entire industries could collapse, exacerbating social unrest or poverty.

Ethical Misalignment

As AI continues to improve, it may develop decision-making processes that humans can't audit. This could lead to morally questionable actions, such as prioritizing certain groups or outcomes based on unintended biases embedded in its initial programming or data.

Power Concentration

If only a few entities control such AI systems, they could wield disproportionate power, leading to geopolitical instability or authoritarianism. Conversely, if the technology spreads uncontrollably, malicious actors could exploit it for harmful purposes, like creating autonomous weapons.

Cultural and Psychological Impacts

Widespread reliance on super-intelligent systems may erode human agency, creativity, and critical thinking. People could become overly dependent on AI, diminishing their ability to solve problems independently or maintain meaningful social structures. For example, we used to have everyone's phone numbers memorized, but now we have all those numbers stored on our smartphones, and don't remember phone numbers at all. We will likely see the greatest harm to the most vulnerable, the children growing up in the AI age. We'll dive deeper into this in the next chapter.

Unpredictable Feedback Loops

Recursive self-improvement could lead to unforeseen interactions with other technologies or global systems. For example, an AI optimizing financial markets might trigger economic chaos through rapid, opaque trading strategies. While these risks are speculative. They underscore the need for robust safeguards, including ensuring that AI goals remain aligned with human values through ongoing oversight. Implementing kill switches and containment to halt or limit AI operations if they deviate from their intended purpose.

Therefore, we need transparent development to prevent monopolistic control and hidden biases, as well as regulatory frameworks to establish global standards for managing AI development and deployment. The implementation and governance of recursive self-improvement determine whether these consequences occur. The challenge lies in balancing innovation and caution to avoid outcomes that could destabilize or harm humanity.

Autonomous Weapons

Consider the development of autonomous drones and weapon systems. Military-grade drones, capable of selecting and striking targets without human oversight, are currently being tested in conflict zones. These systems, driven by AI, can process data faster than any human, and they can save lives on the battlefield. However, since they lack moral judgment, a glitch or misjudgment could escalate a skirmish into a catastrophe. The risk isn't just technical; it's ethical. Who's responsible when a machine kills?

Companies like Palantir and Anduril are leading the charge in defense tech, harnessing AI to power autonomous weapons and reshape modern warfare. Palantir's data analytics platforms, such as Maven, process vast amounts of battlefield data to enable rapid decision-making. At the same time, Anduril's Lattice OS drives autonomous drones and surveillance systems, producing thousands of units to outpace traditional defense giants. These firms, backed by billions in venture capital, are innovating and redefining national security with AI-driven precision.

To stay ahead of these challenges, we should advocate for strict regulations. As citizens, we have the power to support these efforts by staying informed and educating policymakers to act. Upskilling in ethics or policy analysis will equip you to contribute to these conversations and debates, ensuring AI serves humanity, not harm. The actions we take today matter, and they can make a difference.

Living in Virtual Reality Simulations

Another risk associated is the allure of virtual reality (VR) simulations, which are increasingly powered by artificial intelligence. Companies like Meta have poured billions into the metaverse, creating immersive digital worlds where people work, socialize, and even "live."

The danger lies in over reliance. If we outsource our social lives or sense of purpose to virtual worlds, we risk losing touch with what makes us human: connection, struggle, and growth. AI-driven simulations can blur the line between reality and escape, where people could become trapped in a matrix environment, making them less social and more dependent on technology.

To counter this, we should prioritize real-world skills such as communication and community building. Examples include joining a local volunteer group or learning a physical craft, such as woodworking, which grounds you in tangible reality. A balanced approach is key. VR should be used as a tool for creativity and training, not as a substitute for life.

Peter & Misty Phillip

Chapter 9
Protecting the Digital
Generation

We are raising the first generation of children who are interacting with artificial intelligence, but we're still learning what that means for their minds, their safety, and their future. A mother watched her eight-year-old daughter swipe through yet another video on her tablet, eyes glazed with a familiar hypnotic stare. Minutes earlier, the child had been watching educational content about marine biology. Now, a makeup tutorial clearly designed for teenagers absorbed her. The algorithm had seamlessly guided her from learning about dolphins to learning about lip gloss, and the mother felt a chill of recognition. This wasn't just entertainment. This was behavioral programming happening in real time, and it is happening every day in homes all around the world.

Across the country, millions of parents are witnessing similar scenes, often without realizing they're watching the most sophisticated influence campaign in human history unfold on their kitchen tables.

The Attention Economy's Youngest Victims

The battle for human attention has found its most vulnerable targets: developing minds that lack the cognitive defenses to resist sophisticated manipulation. Unlike traditional media that delivered content in scheduled blocks, modern systems continuously adapt to maximize engagement, creating what researchers call "digital quicksand". The more children struggle to look away, the deeper they sink.

Consider the mechanics at work. Every tap, swipe, pause, and replay generates data points that feed back into recommendation engines designed by teams of neuroscientists, behavioral economists, and data scientists. These professionals aren't trying to harm children, but they're optimizing for metrics that often conflict with healthy development.

The pharmaceutical industry learned this lesson decades ago when marketing to children. Regulations now restrict how drug companies can advertise to minors, recognizing that children can't make informed medical decisions. Yet we allow technology companies to conduct unsupervised psychological experiments on developing brains, often without parents even understanding what's happening.

The Invisible Curriculum

Every interaction with these systems teaches lessons that extend far beyond the apparent content. A child learning math through a gamified app isn't just absorbing multiplication tables; they're learning that education should be constantly entertaining, that focus requires external stimulation, and that learning happens in bite-sized, instantly rewarding chunks.

Implicit lessons such as these shape expectations for all future learning. Teachers report increasing difficulty maintaining student attention during activities that don't provide immediate feedback or entertainment value. Students struggle with books, lectures, and complex problem-solving that requires sustained concentration. The invisible curriculum of instant gratification is undermining the visible curriculum of traditional education.

The Trust Trap

Children naturally assume that information delivered through sophisticated interfaces must be accurate and trustworthy. Biased, incomplete, or deliberately misleading datasets used to train systems that generate or curate content make this assumption dangerous. A child asking about a sensitive historical topic might receive an answer that's technically accurate but developmentally inappropriate, or worse, subtly biased in ways that shape their worldview without their awareness.

The problem compounds when children prefer these interactions over human ones. Digital teachers never get frustrated, never have bad days, and always provide immediate responses. They seem perfect until you realize that perfection in teaching often comes from imperfection, from the human ability to struggle alongside students and model how to work through confusion and uncertainty.

The Manipulation Machine

The line between education and exploitation has blurred beyond recognition. Children encounter advertising disguised as content, product placements embedded in games, and social pressure manufactured by algorithms designed to maximize engagement and spending.

Consider the ecosystem surrounding popular gaming platforms. Children watch influencers play games, receive recommendations for similar content, see advertisements for game-related products, and face peer pressure to purchase in-game items or accessories. The entire experience is orchestrated to funnel them toward spending decisions they're not developmentally equipped to make.

The manipulation extends beyond commerce into social dynamics. Platforms amplify content that generates strong emotional responses, from outrage to envy and excitement, because these emotions drive engagement. Children who are still learning to regulate their emotions become unwitting participants in systems designed to keep them in heightened emotional states.

The Privacy Paradox

Children's digital interactions generate unprecedented amounts of personal data, creating detailed psychological profiles that will follow them into adulthood. Every search query, every pause while reading, every game choice, and every social interaction feeds into databases that know these children better than they know themselves.

This data collection happens largely invisibly. Parents struggling to understand privacy policies written by lawyers for lawyers can't meaningfully consent to data practices they don't understand, and children certainly can't. The result is a generation growing up under surveillance they're unaware of, with implications that won't become clear for decades.

The stakes extend beyond privacy violations into questions of human autonomy and development. When systems can predict a child's behavior better than the child can, when they can manipulate emotions and decisions with increasing precision, what happens to the development of independent judgment and critical thinking?

The Security Illusion

The smart toys, educational apps, and connected devices that populate modern childhoods create new attack vectors for those who would exploit children. Traditional safety measures focused on physical proximity, like don't talk to strangers, don't go places alone. Digital safety requires understanding threat models that many adults find incomprehensible.

The problem isn't just external threats. The devices given to children often collect and transmit data in ways that parents don't understand or expect. A tablet designed for children might record conversations, track locations, or monitor app usage in ways that violate reasonable expectations of privacy.

Building Digital Wisdom

The solution isn't to eliminate technology from our lives and the lives of our children. That would be both impossible and counterproductive. Instead, we need to develop what researchers call "digital wisdom". This wisdom is the ability to use technology in ways that enhance rather than diminish human flourishing.

This requires rethinking our approach to both technology design and child development. Instead of systems optimized purely for engagement, we need platforms designed with child development principles in mind. Instead of assuming that more screen time equals better preparation for a digital future, we need to recognize that developing concentration, empathy, and critical thinking might require less screen time, not more.

Progressive schools are already experimenting with approaches that integrate technology thoughtfully rather than reflexively. Students learn to code not just as consumers of technology but as creators who understand how systems work and can make informed choices about when and how to use them.

The Regulatory Awakening

Governments worldwide recognize that children's digital experiences require special protection. However, regulation alone can't solve the problem. Technology evolves faster than law, and enforcement across global platforms remains challenging. The most effective protection will come from a combination of thoughtful regulation, responsible corporate behavior, informed parental choices, and digital literacy education.

Preparing the Next Generation

The children growing up today will shape the digital future in ways we can't predict. Whether they become thoughtful creators and critical consumers of technology, or passive victims of increasingly sophisticated manipulation systems, depends on the choices we make now.

This means teaching children not just how to use technology, but how to think about it critically. It means helping kids to understand that systems designed to capture their attention may not have their best interests in mind. It means modeling healthy relationships with technology and creating spaces for the kinds of deep thinking and sustained attention that digital systems often undermine.

The stakes couldn't be higher. We're not just protecting individual children; we're shaping the cognitive and emotional capabilities of the generation that will inherit and shape our digital future. The choices we make today about how children interact with these systems will determine whether technology amplifies human potential or diminishes it.

The question isn't whether children will grow up with technology; they will continue to do so. The question is whether we'll design the relationship thoughtfully or let it develop by accident, guided by the profit motives of corporations rather than the developmental needs of children.

Our children deserve better than to be the unwitting subjects of the largest uncontrolled experiment in human development in history. They deserve a digital environment designed for their flourishing, not just their engagement. Creating that environment requires all of us to understand what healthy child development looks like in the digital age.

Part II: Strategy

Peter & Misty Phillip

Chapter 10
The Coming Shift

There is a coming shift in the not too distant future, and it will radically change the way we live and work. For millions of people around the world, the alarm clock goes off at 6:30 AM, just as it has for decades. These people shower, grab coffee, and head to work, and many are unaware that they may be among the last generation to follow this routine. While they sleep, artificial intelligence systems are learning to do their jobs faster, cheaper, and often better than they ever could.

This isn't a distant-future scenario. The Great Disruption is already upon us, and its impact is accelerating. We wrote this book to help prepare you for the days of coming disruptions.

The Velocity of Change

Please understand, we stand at an inflection point unlike any in human history. Previous industrial revolutions unfolded over decades, giving workers time to adapt and economies time to adjust. The steam engine took nearly a century to reshape manufacturing. The personal computer required thirty years to transform office work. But artificial intelligence and automation are compressing these timelines into mere years, sometimes months, and even weeks.

Consider this scenario. ChatGPT reached 100 million users faster than any technology in history. Meanwhile, generative AI capabilities that seemed impossible in 2020 became commonplace by 2023. The pace isn't just speeding up; it's exponential. And with each breakthrough, entire categories of human work will become obsolete overnight. We cannot predict this impact on humanity. This rapid pace of transition underscores the need for continuous learning and adaptation.

In the United States alone, one-third of all jobs, representing 50 million workers, face a significant risk of automation within the next decade. These are no longer just factory jobs. Lawyers, accountants, radiologists, and even creative professionals are watching as algorithms encroach on territories once thought to be uniquely human. Jeopardizing almost every profession you can think of.

Technology has been disrupting jobs for ages. Examining the past provides insight into what's happening with AI today. Let's take a quick trip through some jobs that got turned upside down by advancements. Word processors replaced typists, automated exchanges eliminated telephone operators, digital projectors sidelined film projectionists, online booking platforms displaced travel agents, and ATMs and apps reduced the need for bank tellers. But the AI revolution is a different beast, accelerating at breakneck speed and reshaping entire industries with unprecedented scope, from automating complex decision-making to generating creative content in seconds.

These transitions are proof that change is nothing new, but it sure keeps us on our toes! From typists and secretaries to telephone operators and film projectionists, the impact of technology on these roles indicates the transformative power of innovation. This rapid transformation demands that workers and businesses adapt swiftly or risk obsolescence, making lifelong learning and agility crucial for staying relevant in an AI-driven world.

The Hierarchy of Vulnerability

Not all jobs face equal risk in the coming disruption. Understanding where you stand on the vulnerability spectrum, which we refer to as 'The Hierarchy of Vulnerability', could mean the difference between thriving or becoming obsolete. This concept categorizes jobs based on their susceptibility to automation, with some roles at the top facing immediate danger and others at the bottom having a lower risk of displacement.

The Immediate Danger Zone

Manufacturing workers have been living this reality for years, but the pace is about to intensify dramatically. Advanced robotics can now perform complex assembly tasks that once required human dexterity and judgment. Tesla's factories already operate with minimal human intervention, and Amazon's warehouses increasingly rely on robots for picking, packing, and sorting.

But it's not just traditional blue-collar work at risk. Construction workers face threats from 3D printing of building components and automated masonry systems. Even skilled trades like plumbing and electrical work are being enhanced by AI-powered diagnostic tools that can identify problems and guide repairs with minimal human intervention.

The transportation sector faces perhaps the most dramatic upheaval. As autonomous vehicles advance rapidly, the 3.5 million Americans who drive trucks for a living, along with taxi drivers, delivery workers, and ride-share operators, are facing an uncertain future. When self-driving vehicles achieve full autonomy, these jobs won't just shrink; they'll disappear entirely.

White-Collar Jobs: The Surprising Target

The most shocking aspect of the Great Disruption is how it's targeting jobs that were once considered safe, including those of white-collar professionals. For decades, parents have told their children to "get an education" and pursue professional careers. Today, those very careers are among the most vulnerable.

Legal professionals discover AI can review contracts, conduct legal research, and even draft basic documents with stunning accuracy. Junior associates at major law firms are finding their roles eliminated as AI handles document review and case research. Financial services face similar disruption. Robo-advisors are replacing financial planners, algorithmic trading has eliminated floor traders, and AI-powered credit decisions are making loan officers obsolete.

Accounting faces an existential crisis too. Tax preparation software has already automated many basic accounting tasks, but AI can perform bookkeeping, financial analysis, and even complex audit procedures. The Big Four accounting firms are quietly automating processes that once employed thousands of junior accountants and auditors.

Healthcare professionals, who have long been immune to automation, find that AI is encroaching on their expertise. Radiologists compete with AI systems that can detect cancer more accurately than human doctors. IBM's Watson can diagnose certain conditions faster than specialists. Even nursing, with its emphasis on human care, faces challenges from AI-powered monitoring systems and robotic assistants.

The Retail Apocalypse Accelerates

Retail has been transforming for years, but AI and automation are speeding up the process to an unprecedented extent. Self-checkout systems are just the beginning. Amazon Go stores operate entirely without cashiers, using AI to track purchases automatically. Other retailers are rapidly deploying similar technologies.

However, the disruption extends beyond checkout counters. AI-powered inventory management systems predict demand with unprecedented accuracy, reducing the need for human buyers and inventory specialists. Customer service is increasingly relying on chatbots and AI assistants that can handle complex inquiries without requiring human intervention. As e-commerce continues to grow and AI automates remaining functions, traditional retail jobs will continue to disappear at an accelerating pace. Currently, Google is reshaping the future of shopping through several innovative AI-powered shopping experiences that personalize product recommendations.

The Creative Paradox

Perhaps most surprising is the threat to creative professions. For generations, people considered creativity a uniquely human trait. Today, AI systems generate art, write articles, compose music, and even create video content that rivals human output. Veo 3, an advanced AI video generator created by Google, creates high-quality photorealistic video with synchronized dialogue, sound effects, and ambient noise from text or image prompts.

Graphic designers watch AI tools create logos and marketing materials in minutes. Copywriters compete with AI that can produce compelling marketing copy, blog posts, and even books. Musicians see AI composing original songs in any style. The creative industries, which once seemed immune to automation, discover that pattern recognition and recombination underlie much of what we commonly refer to as creativity.

The Deceptive Safety of "People-Facing" Roles

Many people take comfort in roles that require human interaction, believing these jobs are safe from automation. This confidence may be misplaced. AI-powered chatbots and virtual assistants are becoming increasingly sophisticated in handling customer interactions. They work 24/7, never have bad days, and can access vast databases of information instantly.

Sales representatives face AI systems that can analyze customer behavior, predict purchasing decisions, and personalize pitches with greater precision than human salespeople. Customer service representatives compete with AI assistants, who can resolve most issues without human intervention. Even teachers confront AI tutoring systems that can personalize instruction for individual students. This is all happening faster than the average person can even imagine.

The Cognitive Threat

These examples show how technology can transform work in an instant, from typewriters to ATMs. The good news is that with each shift there are new open doors. Think software developers or digital marketers. As AI shakes things up today, it's a reminder to stay curious and continue learning.

This AI wave isn't like past tech changes. It's a whole new ballgame! Previously, machines took over physical jobs, such as factory work. Now, AI is tackling brainy tasks, including thinking, analyzing, and decision-making. Things that we thought only humans could do.

AI is super good at spotting patterns, crunching data, and making smart predictions. It can sift through tons of info and catch things we'd miss. This puts jobs like analyzing reports or making big calls at risk. It's not just a single job, either. Entire fields are feeling the heat. Management consultants who solve business puzzles now compete with AI that delivers faster insights.

Market researchers who study what customers want face AI that tracks trends in real time. This is a wake-up call to get ready for change!

The Acceleration Factor

Several forces are speeding up the pace of job displacement beyond historical precedent. First, the marginal cost of deploying AI approaches zero once systems are developed. Unlike previous technologies, which required significant infrastructure investment for each new application, AI developments scale instantly across global operations.

Second, the COVID-19 pandemic accelerated automation adoption by years, if not decades. Companies operating with reduced workforces discovered they could automate or eliminate many functions. Remote work normalized digital interfaces, making it easier to replace human workers with AI systems.

Third, economic pressures are intensifying the drive toward automation. Increased labor costs, supply chain disruptions, and competition are pushing companies to find more predictable and scalable solutions. AI offers the promise of consistent performance with no sick days, vacation time, or labor disputes.

The Compounding Effect

Job displacement creates a compounding effect that speeds up further disruption. As AI systems replace workers in one area, those displaced workers often move to other sectors, increasing competition and driving down wages. This economic pressure makes automation more attractive to employers, creating a cycle that speeds the overall process.

As more companies adopt AI solutions, competitive pressure forces others to follow suit or risk being outpaced by more efficient competitors. This creates industry-wide shifts that can happen remarkably quickly, as we've seen in sectors such as photography with digital cameras, the music industry with streaming, and retail with e-commerce.

The Skills That Matter

With the AI revolutionizing the landscape, some skills are becoming increasingly valuable while others are fading into obscurity. Strategy and managing AI systems are in hot demand. Human skills, such as creative problem-solving, adaptability, understanding emotions, clear communication, and making ethical choices, are just as important and challenging for AI to replicate. Focus on these, and you'll stand out in this new world!

The workers who will thrive in the post-disruption economy are those who can work alongside AI systems, leveraging artificial intelligence to amplify their human capabilities rather than competing with machines on their terms. This requires a fundamental shift in how we think about work and skills development. The key is developing skills that complement AI rather than compete with it directly.

The Urgency of Now

The window for adaptation is narrowing rapidly! Unlike previous industrial transitions that unfolded over generations, the Great Disruption demands immediate action. Workers who wait for clear signals about which jobs will survive may find themselves too late to develop the skills needed for the new economy.

The pace of change means that traditional education and training systems, which were designed for slower transitions, are now inadequate for current needs. Waiting for institutions to adapt means accepting obsolescence. Individual workers must take responsibility for their transformation, developing new skills and capabilities before their current roles become obsolete.

This urgency applies not just to individuals, but to entire organizations and communities. Companies that delay automation strategies risk being out-competed by more agile competitors. Communities that cannot prepare their workforces face economic devastation as major employers automate or move operations.

The Great Disruption isn't coming; it's here. The question isn't whether your job will be affected but when and how severely. Those who recognize this reality and act decisively will find opportunities in the chaos. Those who deny or delay will find themselves swept away by forces beyond their control. The choice is stark but simple: upskill or risk becoming irrelevant. Comfortable assumptions about job security have ended, and the time for transformation has begun.

Chapter 11
A Launchpad for Your Career

Up to this point, we've discussed AI and potential risks associated extensively, including its impending disruptions, but in this chapter, we are going to look at AI as an opportunity, a launchpad for new careers, and new ways of working. While AI may seem intimidating, it's also opening doors to incredible advancements and new career opportunities. AI is rapidly shaking things up in the marketplace, from automating repetitive tasks to providing enhanced data analysis and creating personalized customer experiences.

While the Great Disruption threatens millions of jobs, it's simultaneously creating unprecedented opportunities for those willing to embrace change. In the not too distant future, there will be a growing demand for AI engineers, machine learning specialists, and data engineers to build, train, and maintain AI models, particularly in industries like healthcare, retail, and manufacturing. As AI adoption grows, roles like AI ethicists, compliance officers, and bias auditors will emerge to ensure responsible AI use, addressing fairness, transparency, and regulatory compliance. Prompt engineers design and optimize the text inputs given to AI systems to produce more accurate, useful, and reliable outputs for specific tasks or applications. The same technologies that eliminate traditional roles are generating entirely new careers, industries, and ways of working that didn't exist five years ago.

The Hidden Truth About Creative Destruction

Innovation has a way of destroying old ways of doing things while creating new opportunities. From airplanes to automobiles, from personal computers to smartphones and the Internet are all examples. As we've already shown, the personal computer eliminated thousands of typing pool jobs but made millions of new roles in software development, digital design, and information management. The Internet has disrupted traditional retail and media models, giving rise to e-commerce, social media, and the gig economy.

The rise of artificial intelligence is causing big changes, but it's unique. The scale and speed of opportunity creation are unprecedented. For every job category that AI threatens, it's creating multiple new roles that leverage human-AI collaboration. The key is to recognize these opportunities early and position yourself accordingly to capture them.

The AI economy is not just a shift; it's a revolution that's creating job categories that sound like science fiction but offer very real career paths today. These roles fall into several key areas. Let's dive in and explore this exciting new career landscape.

AI Development and Management

The most obvious opportunities involve building and managing AI systems. But these roles extend far beyond traditional programming. Human-AI specialists improve communication with AI. AI Product Managers guide the development of intelligent systems without writing code, while AI Ethics Officers ensure the responsible deployment of AI technologies. Prompt engineers craft high-performance AI prompts and now command six-figure salaries, crafting effective instructions for AI systems.

Machine learning operations (MLOps) engineers bridge the gap between AI research and practical implementation. Conversational AI designers create the personality and interaction patterns for chatbots and virtual assistants. AI training data specialists curate and prepare the information that teaches AI systems to function effectively.

Tools like generative AI assist in content creation, design, and innovation, leading to roles for creative technologists who blend AI tools with human creativity in marketing, media, and product design. AI optimizes production and supply chains, increasing demand for roles like automation engineers, predictive maintenance specialists, and logistics optimizers. AI also improves threat detection and response, creating opportunities for cybersecurity analysts and AI security specialists to develop and monitor AI-driven defense systems.

Human-AI Collaboration Roles

Perhaps more significant are jobs that combine human judgment with AI capabilities. Augmented analytics specialists utilize AI tools to uncover insights in data but rely on human expertise to interpret the implications and recommend actions. AI-assisted designers use generative AI to rapidly generate concepts, then refine the results by applying human creativity and aesthetic judgment. New jobs like AI trainers, who refine AI models, and human-in-the-loop moderators, who oversee AI decisions, will emerge to ensure effective collaboration.

Healthcare is seeing explosive growth in AI-augmented roles. Medical AI interpreters help doctors understand and trust AI diagnostic recommendations. Precision medicine coordinators use AI to personalize treatment plans while managing the human elements of patient care. Robotic surgery technicians work alongside surgical robots, handling the complex human factors that machines cannot operate. AI supports diagnostics, drug discovery, and patient care, creating opportunities for medical data analysts, AI-healthcare integrators, and Telehealth specialists.

The Trust and Safety Economy

As AI becomes more prevalent, society needs people to ensure these systems work fairly and safely. AI auditors examine algorithms for bias and effectiveness. Algorithmic accountability managers assist organizations in complying with emerging AI regulations. Digital rights advocates protect individuals from AI-driven discrimination and privacy violations.

Deepfake detection specialists identify synthetic media and combat misinformation. AI safety researchers work to prevent the unintended consequences of advanced AI systems. These roles combine technical understanding with ethical reasoning and social awareness, and uniquely human capabilities that will only grow in importance in the AI economy.

New Industries and Business Models

AI is creating entirely new industries that require human workers at every level. The drone economy, powered by AI navigation and analysis, needs pilots, maintenance technicians, data analysts, and regulatory specialists. Autonomous vehicle deployment requires safety drivers, remote monitors, fleet managers, and customer experience specialists.

Virtual and augmented reality, enhanced by AI, are creating careers in immersive experience design, virtual world architecture, and digital twin management. The creator economy benefits from AI tools that help individuals produce content more efficiently, leading to new roles in AI-assisted content creation, digital marketing automation, and personalized education delivery.

Breaking Down the Barriers

One of the most damaging myths about the AI economy is that it's only accessible to young people with computer science degrees or technocrats. The reality is far more encouraging, with the most valuable AI-era skills often being built upon existing knowledge and experience rather than replacing them entirely.

Contrary to popular belief, older workers often have significant advantages in the AI economy. Years of industry experience provide the context and judgment that AI systems lack. Understanding how businesses operate, what customers really want, and where processes typically break down becomes more valuable, not less, in an AI-driven world.

Background Diversity Drives Innovation

The AI revolution is not just about technology; it's about people. It rewards diverse perspectives and experiences. Former artists become prompt engineers for generative AI, blending their creative talents with AI efficiency. Ex-military personnel will excel at AI security and risk management. Retail workers are moving into AI customer experience design. The key insight is that AI amplifies human expertise, and your unique background could be the next driving force of innovation. Roles such as AI-assisted urban planners who design cities with autonomous traffic flow, or digital legacy managers who curate our AI-generated data after we are gone.

The Learning Revolution

The traditional model of education, four years of college followed by forty years of work, is obsolete in the AI era. Instead, continuous learning becomes a career requirement. But don't be discouraged; new tools and approaches make this more accessible than ever, and your commitment to education will be the key to your success in the AI economy.

Micro-Credentials and Skill Stacking

Rather than pursuing entire degree programs, workers can now build careers through skill accumulation, also known as skill stacking. Much like habit stacking, skill stacking is combining specific competencies that create unique value propositions. A warehouse worker might add drone operation certification, data analysis skills, and supply chain knowledge to become a Logistics AI Specialist.

Online platforms offer micro-credentials in everything from AI prompt engineering to ethical AI implementation. These focused programs, often completed in weeks or months instead of years, allow workers to develop relevant skills while continuing to earn an income in their current roles.

AI-Powered Learning

Ironically, AI itself is making upskilling more effective and accessible. Personalized learning platforms can adapt to individual learning styles and paces. AI tutors provide instant feedback and support. Language learning apps utilize AI to tailor instruction to the specific needs of professionals.

Virtual reality training programs powered by AI enable workers to practice new skills in a risk-free environment. A potential drone operator can practice hundreds of flight scenarios without risking damage to actual equipment. Healthcare workers can interact with AI diagnostic systems in simulated environments before working with actual patients.

The Mindset Shift

The most crucial element of success in the AI economy is adopting the right mindset about the relationship between humans and artificial intelligence. This requires a willingness to learn and improve. Believing that you can improve your skills through hard work is having a growth mindset.

In the AI era, this mindset becomes essential because the pace of change demands continuous learning and adaptation. It also requires embracing uncertainty and experimentation, with resilience and determination. Traditional career paths once followed predictable trajectories, but in the AI era, careers are developing more dynamically, with new opportunities emerging regularly and requiring workers to pivot and adapt more frequently. Rather than viewing AI as a competitor to be feared, successful career transitioners see it as a powerful tool to be leveraged. Will talk about this more in the next chapter.

The Compound Effect of Early Action

The AI economy disproportionately rewards early adopters. Workers who develop AI-relevant skills while these fields are still emerging often find themselves in leadership positions as industries mature. The accountant who learns AI-powered analysis tools early becomes an AI accounting specialist. A teacher who masters educational AI platforms becomes an EdTech Integration Consultant.

This creates a compound effect, where early investments in AI-related skills generate increasing returns. Workers who wait until AI transformation is obvious compete with many others for entry-level positions in established fields.

Beyond Individual Success

The opportunity presented by AI extends beyond individual career advancement to societal transformation. Workers who successfully transition to AI-enhanced roles become bridges between traditional industries and technological innovation, serving as a vital link between the two. They help ensure that AI implementation considers human needs, ethical implications, and practical realities.

The human element becomes increasingly valuable as AI capabilities continue to expand. Society needs people who understand both the potential and the limitations of AI, who can guide its development and deployment in beneficial directions, and who can help others navigate the transition.

The Great Disruption doesn't have to mean significant displacement. For those willing to learn, adapt, and embrace change, it represents one of the most incredible career opportunities in human history. The question isn't whether AI will transform your industry, because it will. The question is whether you'll be a passive victim of that transformation or an active architect of your AI-enhanced future.

The tools are available, and the opportunities are emerging. The choice is yours. It is time to begin now!

Real people are already leaping. Truck drivers now fly drones for farmers, using their road knowledge in a high-tech role. Burned-out teachers are designing AI-powered learning tools to provide more significant help to children. Retail managers become AI marketing strategists, turning customer skills into e-commerce magic. What is their secret? They saw AI as a tool, not a threat, and learned new skills to work with it.

You don't need to be young or a tech genius to jump in. Your experience is a goldmine in this AI gold rush. Start by listing what you know, like how your job works or what customers need. Then, learn a bit about AI tools, data basics, or how to team up with AI. Online courses, such as micro-credentials in AI ethics or drone technology, make it easy. Act now, and you could lead the way in this AI-powered future.

Chapter 12
The Upskilling Mindset

Our world is changing at an unprecedented rate, and standing still is no longer an option. To succeed and remain competitive in today's economy, it's essential to adopt an upskilling mindset. But what is upskilling anyway? Upskilling is acquiring new skills or refining existing ones to remain relevant in a rapidly changing job market, particularly in response to technological advancements such as artificial intelligence.

The ability to adapt, learn, and grow continuously is what separates those who thrive from those who merely survive. This is the essence of the upskilling mindset. A powerful approach to personal and professional development hinges on two core principles we will discuss throughout this chapter.

First, embrace a growth mindset, and second, overcome the fear of change. These concepts form the foundation for building a career and life of resilience, opportunity, and impact. In this chapter, we'll explore how to cultivate this mindset, confront the barriers that hold us back, and harness the tools to keep learning and growing in an ever-changing world.

Embracing the Growth Mindset

The belief that effort, persistence, and learning can develop abilities, skills, and intelligence builds the upskilling mindset. This perspective, unlike a fixed mindset, which assumes our talents and capabilities are set in stone, recognizes that we can enhance these qualities. A growth mindset sees difficulties as chances to learn, mistakes as lessons, and effort as the key to expertise.

To adopt this mindset is to view every experience, whether it is a success or a failure, as an opportunity to learn and grow. When our kids were young, we understood that pain is a teacher, and failure would teach them to problem-solve and would make them resilient.

Growth isn't just about acquiring new skills. It's also about embracing a purpose-driven life where learning becomes a tool for both influence and impact. You can transform your career by leaning into discomfort, learning new skills, and using growth to inspire others. This aligns perfectly with the upskilling mindset! It's not enough to learn on your own. Your growth can ripple outward, creating opportunities for those around you.

To cultivate a growth mindset, start with these practical steps:

Reframe Challenges as Opportunities

Instead of becoming frustrated by the challenges of the AI revolution, let's reframe these challenges as opportunities for growth and for gaining new skills. The rise of AI is transforming our world faster than most of us can keep up and at a pace that outpaces most of us. It's easy to feel frustrated by the fear of losing your job, the challenging learning curve of new tech, or just the relentless pace of it all. But what if you looked at these changes differently? What if, instead of feeling stuck or intimidated, you saw them as chances to grow?

That's at the heart of the upskilling mindset. It's not about having all the answers. Instead, it's about being willing to learn, adapt, and turn fear into forward motion. Challenges are invitations to step into your potential. When you stop viewing AI as a threat and start seeing it as a tool, whether that means learning a bit about machine learning, exploring data trends, or understanding how AI is shaping ethics and policy, you open yourself up to new opportunities. And you don't have to do it all at once. Start with one new skill, one course, or one project. Every step you take builds confidence and helps you gain the skills necessary to succeed in the AI revolution.

Celebrate Effort, Not Just Results

When you're navigating unfamiliar territory, whether it's learning AI tools or pivoting your career, it's tempting to measure success only by visible wins. But the real growth happens in the trenches. Growth occurs in the hours you commit to learning and growing, the mistakes you learn from, and the courage it takes to try something new. That kind of effort deserves recognition. Celebrate the fact that you showed up, stayed curious, and didn't give in to fear. Progress isn't always about crossing a finish line; instead, it's about becoming the person who keeps running even when the path isn't clear. When we learn to value effort as much as outcomes, we reinforce the belief that growth is possible, even in uncertain times.

When you celebrate effort, not just results, you can rejoice in the grit it takes to show up, the curiosity that drives you to keep learning, and the bravery to face uncertainty. You cultivate a belief that progress is possible, even when the path ahead may be unclear.

To make this mindset stick, try these steps. First, track your efforts by keeping a journal of your daily learning. Track the time spent, questions asked, or mistakes tackled. Reviewing your progress reminds you how far you've come. Second, ensure that you reward the process by treating yourself after you develop a new skill or achieve a goal. Small rewards reinforce the habit of effort.

Share Your Journey

Share a challenge you're currently facing with a friend or colleague. Sharing your story amplifies, keeps you accountable, and inspires others. Finally, reframe your setbacks. When things don't go as planned, ask, "What did I learn?" instead of "Why did I fail?" This shift turns stumbles into progress.

Overcome the Fear of Change

While a growth mindset lays the foundation for upskilling, the fear of change can be a formidable barrier. Most people I know dislike change because it makes them feel uneasy. Whether it's learning a new tech, switching careers, or adapting to changes in the workplace, this often triggers discomfort, self-doubt, or even paralysis. I recently experienced these feelings as I transitioned from a comfortable career in podcasting and owning a media company to embark on the unknown world of tech. This fear is part of the human experience. We are wired to seek safety and predictability. Yet, in a world where automation, AI, and global competition are reshaping industries overnight, clinging to the familiar is no longer an option. It is a recipe for obsolescence.

It is productive to view change as a catalyst for growth rather than a threat to our existence. Fear of change is inevitable, but it doesn't have to define us and our actions. Instead, it can be a signal that you're on the cusp of something transformative. The fear of change often stems from a fixed mentality, where failure feels like a verdict on your worth. To overcome this fear, we must overcome our own self-imposed limitations and reframe the concept of change.

Common thoughts and mindsets that may hold you back:

"I don't have time": If you have time to scroll through social media, then you have time to incorporate micro-learning. Replace 30 minutes of scrolling with learning. That's 15 hours of skill-building per month. Begin micro-learning in 15-minute to 30-minute blocks or use commute time for audio-based courses, podcasts or audio books.

"I'm too old to learn this": The fastest-growing demographic in coding bootcamps is people over 40. Experience plus new skills equals competitive advantage. You are never too old to learn new skills. If you struggle with learning new technologies enlist the help of your kids or other young people in your life to help you get up to speed.

"It's too complicated": Start with Excel. If you can use Facebook, you can learn basic data analysis. Break it down into small bite-size pieces and build on each skill. You will gain confidence with each small win.

"I'll never be as good as the experts": Don't worry, you don't need to be an expert. You just need to be better than you were yesterday. With every skill you gain, you become more valuable and competitive in the marketplace.

Here are some actionable strategies to help you conquer the fear of change and fuel your upskilling journey.

Start Small and Build Confidence

Divide the change or skill you want to learn into manageable steps. Eat the elephant one bite at a time. If you're learning a new skill, such as data analysis, start with a beginner-friendly course or a single tool, like Excel, before progressing to more advanced platforms like Tableau. Small wins build momentum, help you gain confidence, and help reduce fear.

Visualize the Cost of Inaction

Ask yourself, "What happens if I don't adapt?" The prospect of being left behind, whether in your career or personal growth, can be a powerful motivator. Evaluate the long-term impact of staying stagnant versus the temporary discomfort of change. When it comes to creating a future, you are excited about living in, take action and count the cost to help you build a better tomorrow.

Reframe Failure as Feedback

A mistake isn't the end; it's data. When you feel you've messed up or made a mistake, treat each misstep as a valuable learning opportunity. For instance, if a coding project or marketing plan fails, analyze what worked well, what went wrong and try again. This is how we reframe failure as feedback.

The Upskilling Mindset in Action

The upskilling mindset is more than a buzzword. It's a way of life. It's about viewing every challenge and every disruption as an opportunity to grow stronger and more capable. By combining a growth mindset with strategies to overcome fear, you unlock the ability to adapt to any environment. This mindset isn't just about personal gain; it's about becoming a force for positive change in your workplace, community, and beyond.

Take the story of a teacher who, during the shift to online learning in 2020, faced the daunting task of mastering virtual platforms. Initially overwhelmed, they adopted a growth mindset, viewing the challenge as an opportunity to connect with students in new ways. They took online courses, experimented with tools like Zoom and Google Classroom, and sought feedback from students and peers. Over time, they not only adapted but also became leaders in their schools, training others to navigate the digital landscape. Their story echoes what Misty calls sparking your influence through resilience and learning.

To put the upskilling mindset into action, commit to:

Set Clear Learning Goals

Identify one skill to develop in the next three months, whether it's coding, public speaking, or financial analysis. Break it down into weekly tasks to stay focused. Learn to embrace discomfort. Choose one area where fear of change is holding you back. Take a small step forward, like signing up for a workshop or shadowing an expert. Next, reflect and iterate. At the end of each week, reflect on what you've learned and where you struggled. Use these insights to adjust your approach.

Surround Yourself with Growth-Oriented People

By adopting these habits, you see upskilling not as a one-time event but as a lifelong journey of growth. Every new skill you gain and every challenge you overcome strengthens your ability to adapt, pivot, and thrive. We grow exponentially when we listen to the wisdom of our colleagues, mentors, or peers and muster the courage to act on it. Doing this unlocks exponential progress. So, seek feedback and then act on it. Feedback is more than advice. It is a mirror that shows you where you stand and a map that shows you where you can go.

Build a Support System

Surround yourself with people who encourage you to take calculated risks and who inspire you to grow. We need one another as we navigate the changes brought about by this AI revolution. Building supportive networks will help you navigate fear and change and stay accountable to your goals. Overcome barriers to upskilling, such as lack of time, financial constraints, and learning disabilities, with micro-lessons, free online resources, or mentors who can help guide you.

Consider the example of a marketing professional who feared transitioning to digital marketing as traditional methods became less relevant. Initially, the complexity of SEO, social media algorithms, and analytics felt daunting. By starting with a single online course, seeking mentorship from a colleague, and treating early mistakes as learning opportunities, they gradually built confidence. Within a year, they were leading digital campaigns with skills that not only secured their job but also opened alternative career paths.

Upskill or Die

The upskilling mindset is your lifeline in a rapidly changing world. By embracing a growth mindset, you transform challenges into opportunities and failures into lessons. By overcoming the fear of change, you unlock the courage to step into the unknown and emerge stronger. Together, these principles empower you not only to survive but also to thrive in an ever-changing landscape.

The choice is stark: upskill or die, not in a literal sense, but in the sense of relevance, opportunity, and impact. The world won't wait for you to catch up—so start now. Pick one skill, face one fear, and take one step toward growth. As Spark Your Influence reminds us, your influence begins with your willingness to grow.

Remember, your potential is limitless when you believe you can improve. So, what's your next step?

Chapter 13
How to Pivot When Your Industry Changes

The most dangerous phrase in business is: 'We've always done it this way.'

Peter and I used to spend every Friday night at our local Blockbuster looking for movies to rent for the weekend. When Netflix announced in 2007 that it was shifting from DVD-by-mail to streaming, its stock plummeted by 80%. Industry experts called it corporate suicide. Blockbuster laughed. Reed Hastings and his team faced a choice that every professional confronts today: evolve or become extinct.

Today, Netflix, the company that dared to adapt, is worth over $150 billion. Its success story is a testament to the power of adaptability. Blockbuster, the company that laughed at change, is now a cautionary tale told in business schools.

The difference wasn't just technology—it was adaptability.

The New Reality

Change is not just a possibility; it's our only constant. We're living through the most rapid period of transformation in human history. Entire industries are being born, transformed, or eliminated within months, not decades. Yet most professionals approach change like deer in headlights—frozen, hoping it will pass them by. This is the path to professional irrelevance. Adaptability is not just a choice; it's an urgent necessity in this rapidly changing economy.

The Adaptability Imperative

In today's rapidly transforming economy, adaptability isn't just a competitive edge; it's a professional lifeline. Integrating artificial intelligence into workplaces is reshaping industries, redefining roles, and demanding a workforce that can pivot with agility.

The adaptability imperative encapsulates the urgent need for individuals and organizations to embrace continuous learning, flexibility, and resilience to thrive amidst technological disruption. In this fast-paced business landscape, the ability to learn and adapt continuously is a necessity. Below, we'll expand on this concept, exploring its implications, strategies for implementation, and its transformative potential in the AI-driven workplace.

The Anatomy of a Successful Pivot

If 2020 taught us anything, it is the art of the pivot. Businesses shifted to fully remote or hybrid work models. Professionals adapted to working from home, setting up home offices, learning video conferencing etiquette, and balancing work with family responsibilities. Many restaurants unable to host dine-in customers pivoted to takeout, delivery, and meal kits. Small businesses launched e-commerce sites or used social media marketplaces to sell products online when their physical stores were closed. Doctors and therapists shifted to virtual consultations. These pivots highlight the resilience and creativity sparked by the challenges of 2020.

Based on our personal experience and research into successful career transitions, we've identified three critical phases of effective pivoting. The first phase is recognition and assessment, followed by strategic skill building and then bold execution.

Phase 1: Recognition and Assessment

The first skill of adaptability is pattern recognition. Most people miss industry shifts because they're looking down, focused on their immediate tasks, rather than looking ahead. Successful pivoters develop what we call "environmental scanning." This involves regularly assessing your industry landscape and staying current with the latest trends. In the AI revolution, the landscape is changing daily. Peter has built his successful career in information technology by staying on top of emerging trends and pivoting when he sees the industry changing.

Phase 2: Strategic Skill Building

When we can strategically identify the convergence point between existing strengths and emerging opportunities, we can capitalize on our opportunities. One way Peter has strategically built his skills is by consuming vast amounts of information daily and retaining what he reads. When we started Spark Media, Misty consumed an extensive amount of online content on podcasting, social media, and community building, which she then shared with her membership community.

Phase 3: Bold Execution

The final phase separates the adapters from the wishful thinkers. To succeed, you must have bold execution. Our bold execution in identifying market gaps and filling needs, whether building platforms to empower podcasters and creating a media conference, podcast network, and magazine or developing cybersecurity, AI, and PQC solutions for our clients. Transitioning from corporate careers to new ventures, we've seized opportunities to pursue our passions and to inspire others. Our relentless drive reflects our fearless commitment to taking bold action.

Balancing Technical and Soft Skills: The Hybrid Advantags

Here's where most pivot strategics fail. People focus exclusively on technical skills like learning to code, getting certified in cloud computing, or taking a course in data analysis. Technical skills are necessary but not sufficient.

Professionals thriving through industry changes master what we call the 'Hybrid Advantage'. This strategically combines technical competency and enhanced soft skills. This means not only being proficient in your technical field but also developing skills like emotional intelligence, creative problem-solving, and effective communication. Peter is masterful at this, and his ability to blend technical expertise with these soft skills has been a key factor in his successful career transitions.

The 70-30 Rule

Most successful career pivoters follow a 70-30 skill development approach:

70% of their learning time invested in soft skills (emotional intelligence, creative problem-solving, communication, adaptability itself).

30% focused on targeted technical skills that directly support their pivot.

This seems counterintuitive until you understand the math: technical skills become commoditized quickly, but soft skills compound over time and across contexts.

Creativity as a Competitive Advantage

In the age of artificial intelligence, the ability to be creative is what makes people stand out. While AI excels at processing vast datasets and optimizing tasks, it lacks the nuanced imagination to conceive novel ideas or emotionally resonant solutions. Creative people can develop new and innovative strategies by using both human intuition and the analytical capabilities of AI. AI is great at automating the predictable, but it is human ingenuity and creativity that drive disruption.

Emotional Intelligence

Emotional intelligence is the skill that amplifies every other capability you possess. The ability to read a room is critical. Especially during industry transitions, when organizations are stressed, teams are uncertain, and leaders are making difficult decisions. Professionals who can navigate this emotional complexity while maintaining their contributions become invaluable resources. Your emotional intelligence is a vital asset in your adaptability toolbox. It's a superpower that can guide you through the most challenging transitions.

In the AI revolution, emotional intelligence is a vital competitive edge, enabling humans to complement AI's analytical prowess with empathy, self-awareness, and interpersonal finesse. While AI can process data and automate tasks, it cannot replicate the nuanced understanding of human emotions that drives meaningful connections and decision-making. High emotional intelligence enables individuals to navigate complex social dynamics, foster trust, and develop innovative solutions that resonate emotionally with others. Capabilities AI cannot match. As automation reshapes industries, emotional intelligence ensures humans remain indispensable, driving collaboration and creativity in an AI-dominated landscape.

Creative Problem-Solving

The ability to approach challenges from multiple angles and generate innovative solutions is indispensable in today's rapidly evolving workplace. Creative problem-solving involves breaking free from conventional thinking patterns and embracing calculated risk as an opportunity for breakthrough insights. By consistently reframing obstacles as puzzles to be solved rather than barriers to be overcome, you can cultivate curiosity by asking "what if" questions that others might dismiss. Actively seeking diverse perspectives from colleagues across different departments and industries to solve problems. This approach enables you to identify unconventional solutions that not only address immediate problems but often reveal new opportunities for growth and efficiency. Your willingness to experiment with untested approaches allows you to prototype ideas quickly and iterate based on real-world feedback.

Effective Communication

Master communicators understand that technical expertise is useless unless they translate it into language that resonates with diverse audiences. Effective communication in the hybrid advantage context involves adapting your message to match the technical literacy and priorities of your audience, whether you're explaining complex processes to executives or collaborating with cross-functional teams. Peter has developed the skills of using analogies and storytelling to make technical concepts accessible without oversimplifying them. He listens actively to understand not just what people are saying, but what they need to hear to make informed decisions. This communication ability has allowed him to build trust across organizational levels, advocate for resources and support for his projects, and position himself as a bridge between technical teams and business stakeholders. His ability to articulate the business value of technical solutions has been crucial in securing buy-in for innovative initiatives throughout his career transitions.

The Pivot Playbook: A Step-by-Step Framework

Step 1: Industry Intelligence Gathering

Set up Google Alerts for your industry plus adjacent sectors. Since I was in media, I also set up Google Alerts for my name, and found this to be helpful when people would publish podcast interviews where I was a guest so I could interact with their content. Read posts from key people in your field on LinkedIn and post your own insightful content to attract more followers. Join professional associations within emerging fields, and spend 30 minutes a day reading industry publications. Using AI agents can help as well. ChatGPT can create agents. Just create a new agent, and in plain language tell it what to search for and how often to notify you. It will endlessly perform the task for you and support fairly complex prompts.

Step 2: Skills Gap Analysis

Track and analyze gaps by creating three columns with the headings [Current Skills], [Required Skills for Desired Future], and [Skill Development Priority]. Be brutally honest about your gaps. Remember that soft skills often represent the most significant opportunities for improvement. Get creative in the ways you develop soft skills, such as through cross-disciplinary creative projects, using apps, or creating group challenges that gamify the development of these skills. Consider daily prompts to practice active listening, give constructive feedback, and develop volunteer mentorship roles.

Step 3: The Bridge Strategy

Don't jump directly from your current job to your dream job. First, identify one to two intermediate skills you can develop or positions to apply for that bridge your existing expertise with your desired career path. This reduces risk while building credibility.

Step 4: Network Expansion

Your current network likely reinforces your current industry. When I transitioned from media to tech, I deliberately cultivated relationships within the tech space. I attend conferences, join online communities, and network on social platforms like LinkedIn. We know that relationships often matter more than qualifications during career transitions. It is not always what you know, but who you know. However, don't wait until you're out of a job to tap your network. Instead, regularly connect with your colleagues and support one another.

Step 5: Prototype and Test

Before making dramatic changes, test your pivot through side projects, volunteer work, or consulting assignments. This provides real-world validation while building a portfolio of relevant experience.

Common Pivot Mistakes to Avoid

The All-or-Nothing Trap

Don't quit your job to pursue an entirely new career without first testing the waters. Successful pivots are typically gradual and strategic, not dramatic leaps.

The Technical Skills Obsession

Stop believing that more certifications equal more opportunities. Skills without context, relationships, and soft-skill amplification rarely lead to meaningful pivots.

The Comparison Paralysis

Everyone's pivot looks different because everyone's starting point, constraints, and opportunities are unique. Focus on your strategic path rather than copying someone else's journey.

The Perfectionist's Prison

You don't need to be an expert before you pivot. Competence, adaptability, and the ability to be agile and move quickly always beat perfection.

Your Adaptation Action Plan

Starting tomorrow, implement these three practices:

Weekly Future Scanning: Dedicate time each week to research trends, read about industry changes, and identify emerging opportunities.

Monthly Skills Investment: Allocate time and resources to developing technical skills and soft skills each month. Track progress and adjust based on market feedback.

Quarterly Pivot Planning: Every three months, reassess your industry position, update your skills development plan, and take concrete steps toward your desired future.

Remember: adaptability isn't a destination. It's a capability you develop and strengthen. The people who thrive in a rapidly changing economy aren't necessarily the brightest or most talented. They're the ones who embrace change as an opportunity and develop the hybrid skills needed to navigate uncertainty.

Your industry is changing. Expect your job to change. Your career will require pivots you can't imagine today. The question isn't whether you'll need to adapt—it's whether you'll be ready when the moment arrives

The choice is yours: upskill or die.

Peter & Misty Phillip

Chapter 14
Soft Skills That AI Can't Replace

As artificial intelligence reshapes industries and transforms job markets, a fundamental question emerges: what makes humans irreplaceable? While machines excel at processing data, recognizing patterns, and executing programmed tasks, they struggle with the nuanced, contextual, and deeply human aspects of work that define our most valuable contributions.

Our human experience makes us superior to artificial intelligence.

AI lacks the emotional depth and intuitive judgment that humans bring to ambiguous or interpersonal scenarios, making human soft skills irreplaceable despite occasional missteps. This underscores the unique and irreplaceable role that humans play in the age of AI, making us integral to the future of work.

We are entering a time when artificial intelligence can process data, generate content, and automate complex tasks with unprecedented speed. The human skills that matter most are those that machines cannot replicate or replace. Judgment becomes paramount: the ability to weigh nuanced factors, make ethical decisions under uncertainty and discern when to trust or question AI-generated insights.

Communication skills grow more critical as professionals must translate between human and machine intelligence, articulate complex ideas to diverse stakeholders, and collaborate effectively in hybrid human-AI teams. Critical thinking serves as our intellectual firewall, enabling us to challenge assumptions, identify logical fallacies, and maintain a healthy skepticism about automated recommendations.

Perhaps most important is contextual analysis. The deeply human capacity to understand cultural nuances, read between the lines, and grasp the broader implications of decisions within complex social and business environments remains a uniquely human trait. These cognitive abilities cannot be automated away; instead, they become amplified in importance as they represent the irreplaceable human element in an increasingly AI-driven world, determining who will lead the revolution rather than be displaced by it.

The answer lies not in competing with AI's computational power, but in embracing our distinctly human capabilities that remain beyond the reach of algorithms. These enduring soft skills, including creativity, empathy, critical thinking, leadership, and collaboration, represent our enduring competitive advantage in an increasingly automated world. The emphasis on the enduring nature of these skills should instill a strong sense of confidence in your audience about their indispensable place in the future workforce.

Creativity and the Spark of Innovation

Creativity stands as one of the most distinctly human capabilities. While AI can generate variations on existing patterns or combine elements in novel ways, true creativity involves something more profound. It is the ability to make unexpected connections, challenge fundamental assumptions, and imagine possibilities that don't yet exist.

Human creativity emerges from lived experience, emotional depth, and the capacity for abstract thought that transcends logical parameters. When a designer envisions an entirely new user interface, when an entrepreneur identifies an unmet market need, or when a teacher develops an innovative way to explain complex concepts, they're drawing on uniquely human faculties that combine intuition, experience, and imagination.

Consider how breakthrough innovations often come from unexpected places—the scientist who applies principles from biology to engineering, the artist who sees mathematical beauty in natural patterns, or the manager who borrows conflict resolution techniques from family therapy. These creative leaps require associative thinking and contextual understanding that remain uniquely human.

AI lacks the emotional depth and intuitive judgment that humans bring to ambiguous or interpersonal scenarios, making human soft skills irreplaceable despite occasional missteps. To cultivate creativity in an AI-driven workplace, it's crucial to focus on expanding your exposure to diverse experiences, disciplines, and perspectives. When we travel, engage with art, literature, and fields outside our expertise, it broadens our horizons. But most importantly, it's about continuous learning and adapting to AI. Practice brainstorming without immediate evaluation of feasibility. Embrace failure as a learning opportunity rather than viewing it as a setback. Learning from failure should make you feel more resilient and open to experimentation.

Empathy is The Bridge Between Hearts and Minds

Empathy is the ability to understand and share the feelings of others. Empathy is fundamentally human, a human experience because it requires lived emotional experience and genuine care for others' well-being. While AI can recognize emotional patterns in speech or text, it cannot truly feel or experience emotions in the same way humans do. If Peter and I have a disagreement, AI can't help us solve our relational problems. AI doesn't know all of our shared life experiences. Having empathy is something unique to humans.

In professional settings, empathy manifests as understanding customer frustrations, recognizing when a colleague is struggling, or expecting how changes will affect different stakeholders. This emotional intelligence enables humans to build trust, resolve conflicts, and create inclusive environments where everyone can thrive.

Empathy also drives ethical decision-making in ways that purely logical systems cannot. When faced with complex choices that affect real people, humans can weigh not just efficiency or profit but also fairness, dignity, and long-term consequences for all involved parties.

Developing empathy requires active listening, seeking to understand rather than responding, and regularly putting yourself in others' shoes. For instance, in a customer service role, actively listening to a customer's concerns and understanding their perspective can enhance the service experience. Practice perspective-taking exercises, engage with people from different backgrounds, and make space for emotional considerations in your decision-making processes.

Navigating Complexity and Ambiguity

Critical thinking involves analyzing information objectively, questioning assumptions, and making reasoned judgments despite incomplete or contradictory data. While AI excels at processing vast amounts of information, humans excel at evaluating the quality, relevance, and context of that information.

Human critical thinking encompasses several unique capabilities: recognizing bias and logical fallacies, understanding the limitations of data and models, considering ethical implications, and deciding when facing genuine uncertainty rather than just calculated risk.

In an AI-driven workplace, critical thinking becomes even more valuable as humans must evaluate AI-generated insights, question algorithmic recommendations, and make decisions that require wisdom, not just intelligence. The ability to ask the right questions often matters more than having all the answers.

Strengthen your critical thinking by regularly challenging your assumptions, seeking diverse viewpoints, learning to identify and avoid cognitive biases, and practicing structured problem-solving approaches that consider multiple perspectives and potential consequences.

Leadership in the Age of AI

Leading humans requires understanding their motivations, fears, aspirations, and unique circumstances and areas where human insight significantly exceeds that of artificial intelligence.

Effective leadership in an AI-driven workplace demands several key capabilities that remain distinctly human.

Inspirational Vision

While AI can analyze trends and suggest strategic directions, inspiring others requires the ability to paint a compelling picture of the future that resonates emotionally with people's values and aspirations. Influential leaders tell stories that help people see how their contributions fit into a larger purpose.

This involves understanding what motivates different people, communicating complex ideas in clear and accessible ways, and maintaining optimism and determination even in the face of uncertainty. These capabilities require emotional intelligence, communication skills, and the ability to connect with others on a human level.

Adaptive Decision-Making

Leadership often requires making tough decisions with incomplete information, balancing competing priorities, and adapting quickly to changing circumstances. While AI can provide data and analysis, human leaders must also consider intangible factors, such as team morale, organizational culture, and long-term reputation.

Effective leaders in an AI-driven workplace understand both the capabilities and limitations of artificial intelligence. They know when to trust algorithmic recommendations and when human judgment should take precedence over data-driven suggestions. This requires a combination of technical literacy, wisdom, and experience.

Developing Others

Most importantly, human leaders excel at recognizing potential in others, providing mentorship and coaching, and creating environments where people can grow and develop. This involves understanding individual strengths and weaknesses, providing meaningful feedback, and helping others navigate their career paths.

AI can provide learning resources and track progress. Still, it cannot replace the human connection involved in mentoring relationships or the intuitive understanding of when someone needs encouragement versus challenge.

Cultivating leadership skills in an AI-driven workplace and focus on developing your emotional intelligence. Practice active listening and empathetic communication, learn to decide under uncertainty, and invest time in understanding and developing the people around you.

Collaboration

While AI can facilitate coordination and information sharing, true collaboration requires the human ability to build relationships, navigate interpersonal dynamics, and create a synergy that exceeds the sum of individual contributions.

Effective collaboration depends on building trust and confidence with team members who will act in good faith, support each other's success, and handle conflicts constructively. Building trust requires vulnerability, consistency, and the ability to read and respond appropriately to social cues. Creating this environment requires human leaders who can model vulnerability, encourage diverse perspectives, and respond supportively to mistakes and setbacks.

Every team involves complex interpersonal dynamics influenced by personality differences, communication styles, cultural backgrounds, and individual motivations. Successful collaboration requires the ability to understand these dynamics and adapt your approach accordingly.

This might involve recognizing when someone needs more autonomy versus structure, understanding cultural differences in communication styles, or identifying when interpersonal conflicts are hindering team performance. These insights require emotional intelligence and social awareness that remain uniquely human.

Creative Problem-Solving

When teams collaborate effectively, they generate solutions that no individual could have developed independently. This creative synergy emerges from the interplay of different perspectives, the building upon each other's ideas, and the willingness to explore unexpected directions.

Human collaboration benefits from the ability to improvise, adapt to changing circumstances and pursue promising tangents that emerge during discussions. While AI can suggest combinations of existing ideas, human teams can make creative leaps that transcend logical boundaries.

To enhance collaboration skills, practice active listening and inclusive communication; learn to give and receive constructive feedback; develop conflict resolution skills; and prioritize building genuine relationships with colleagues over just working relationships.

Integrating Human Skills with AI Capabilities

Humans will not define the future workplace versus AI, but humans working alongside AI to achieve results that neither could accomplish alone. This requires developing new competencies that bridge the gap between human and artificial intelligence.

AI Literacy

Understanding AI's capabilities and limitations becomes essential for all professionals. This doesn't require technical expertise, but the ability to ask the right questions about algorithmic recommendations, understand when human oversight is necessary, and recognize potential biases or blind spots in AI systems.

The most effective approach often involves combining AI's analytical power with human judgment and values. This requires knowing when to trust data versus intuition, how to incorporate human considerations into algorithmic recommendations, and when to override AI suggestions based on contextual factors, the system cannot understand.

Continuous Learning and Adaptation

As AI capabilities continue to expand, humans must continually adapt to their roles and develop new skills. This requires intellectual curiosity, comfort with ambiguity, and the resilience to reinvent yourself as needed. Most importantly, it requires maintaining confidence in your unique human value while remaining open to new ways of working.

The human edge in an AI-driven world lies not in competing with machines but in embracing and developing the capabilities that make us uniquely human. Creativity, empathy, critical thinking, leadership, and collaboration represent enduring sources of value that no algorithm can replicate.

By cultivating these soft skills while learning to work effectively alongside AI, we can create a future where technology amplifies human potential rather than replacing it. The organizations and individuals who thrive will be those who understand that our greatest strength lies not in what we can compute, but in what we can imagine, feel, and inspire in others.

The path forward requires the intentional development of these human capabilities, combined with the wisdom to know when to leverage AI and when to rely on distinctly human judgment. In doing so, we preserve not just our economic value but the essential human qualities that make work meaningful and life rich.

Chapter 15
Financial Planning for an AI Future

The economic landscape today is rapidly developing, necessitating a shift in our financial strategies. The methods that served our parents' generation, such as climbing the corporate ladder and relying on pensions, may not suffice in a world where artificial intelligence is reshaping industries overnight. This is not a matter of fear, but of education, adaptation, and positioning. The same technological forces that threaten traditional employment also present unprecedented opportunities for wealth creation.

The New Economic Reality

We're living through an economic transformation as significant as the Industrial Revolution, but compressed into a shorter timeline rather than spanning centuries. Jobs that seemed secure are now being automated, while entirely new professions emerge overnight. The gig economy has already shown that traditional employment models are fracturing, and the acceleration of AI is amplifying this trend.

Consider what happened to photography when digital cameras arrived. One of our friends, Kelly, was a wedding photographer, who pivoted when digital photography transformed the industry. She harnessed her photography skills to become a visual brand strategist and began teaching smartphone photography and teaching people how to take their own headshots with the digital cameras in their pockets. Kelly recognized that while everyone had a camera in their phone, few knew how to use it for professional branding, identifying a direct market need she could fill with her existing expertise. She then further pivoted her offerings to include Canva and AI training as she saw the market shifting.

Streaming services have decimated video rental stores. Podcasts have disrupted traditional broadcasting through radio and television. These changes happened gradually, but cascaded rapidly once the technology reached a tipping point. The problem with the AI revolution is speed and scale. We're approaching similar tipping points across multiple industries simultaneously. We do not know how this will impact us, so we've got to be flexible and have the skills to pivot.

With the AI revolution come both risks and opportunities. The danger lies in assuming that current income sources will remain stable. Opportunity lies in positioning yourself to benefit from the changes rather than being displaced by them.

Building Multiple Income Streams

The era of depending solely on a single employer for financial security is drawing to a close. Smart financial planning now demands diversification not only in your investments but also in your income sources. Rather than relying on a single economic lifeline, think of creating a financial ecosystem. Establishing multiple streams of income can be a key strategy for wealth building.

Start by auditing your current skills and assets. What knowledge do you possess others would pay for, and what problems can you solve? List the assets you possess that could generate income. The goal is to identify multiple pathways to revenue that aren't dependent on traditional employment structures.

Freelancing and consulting represent obvious starting points, but they're still trading time for money. More interesting opportunities emerge when you can create systems that generate income without requiring your constant presence. This might involve creating digital products, building online communities, developing software tools, or investing in assets that appreciate or generate cash flow.

The key insight is that technology, including AI, can amplify your efforts rather than replace them. Instead of competing with automation, you want to leverage it to multiply your output and reach.

Leveraging Technology for Wealth Creation

Modern technology tools can dramatically speed up wealth-building activities that previously required significant capital or specialized knowledge. Online platforms have democratized access to global markets, while sophisticated analytics tools are available to individual investors.

Consider how you can utilize automation to enhance your investment strategies. Automated rebalancing tools can help maintain optimal portfolio allocations without constant monitoring. Algorithmic trading platforms can execute strategies based on predetermined criteria. Tax-loss harvesting software can optimize your tax efficiency automatically.

Beyond traditional investing, technology enables entirely new wealth-creation strategies. You can build and monetize online audiences, create digital products that scale globally, or develop software solutions for niche markets. These approaches use technology to multiply individual effort and yield significant results.

Democratization of content creation tools means you can produce professional-quality materials without expensive equipment or technical expertise. The global reach of internet platforms enables you to access customers worldwide without relying on traditional distribution channels. Sophisticated analytics allow you to optimize strategies using real data instead of guesswork.

Strategic Saving In an Uncertain World

Traditional saving advice often assumes stable employment and predictable inflation. When rapid technological change creates both opportunities and uncertainties, your saving strategies need to adapt accordingly.

Emergency funds become even more crucial when job security is uncertain. However, the traditional advice of keeping three to six months of expenses in low-yield savings accounts may not be sufficient. Consider building larger emergency reserves and ensuring some portion is quickly accessible for time-sensitive opportunities.

This might mean maintaining higher cash reserves than conventional wisdom suggests, but also investing in assets that can be liquidated relatively quickly if needed. The goal is to strike a balance between security and flexibility.

Think beyond traditional savings vehicles. High-yield savings accounts, money market funds, and short-term bond funds can provide better returns than traditional savings accounts while maintaining liquidity. Treasury bills and CDs can provide guaranteed returns for specific time horizons.

Consider also building reserves in different forms. Cash provides immediate liquidity, but holding reserves in appreciating assets, such as stocks or real estate, can offer growth while still being accessible if needed. The key is balancing immediate availability and long-term growth potential.

Investment Strategies for Rapid Change

Investment approaches designed for stable, predictable markets may not perform well during periods of rapid technological disruption. Traditional diversification and spreading investments across different asset classes remains essential, but may not be sufficient when entire industries can be disrupted simultaneously.

Consider tilting your portfolio toward companies and sectors that are positioned to benefit from technological advancements rather than being displaced by them. This doesn't mean abandoning diversification, but being thoughtful about which areas offer the best risk-adjusted returns in a rapidly changing environment.

Technology companies often trade at premium valuations, but some may justify higher prices through superior growth prospects. Similarly, companies in traditional industries that successfully adapt to technological change may outperform those that resist or ignore it.

International diversification becomes important when domestic markets face disruption. Different countries and regions may experience technological adoption at different rates, creating opportunities for geographic diversification.

Real estate can provide both inflation protection and income generation. Still, location and property type are crucial in an economy where remote work is becoming increasingly common, and retail patterns are shifting.

Alternative investments, from commodities to cryptocurrency and private equity, may play a larger role in portfolios designed for uncertain times. However, these require a more sophisticated understanding and typically involve higher risks.

Creating Recession-Resistant Income

Economic downturns are inevitable, but their impact on your finances depends mainly on how you've structured your income and investments. The goal isn't to predict recessions but to build financial resilience that can weather various economic conditions.

Diversified income streams provide natural protection against economic downturns. If one source of income disappears, others can continue generating cash flow. This might include income from investments, royalties from creative work, revenue from online businesses, or income from services that remain in demand, regardless of economic conditions.

Some types of income are naturally more recession-resistant than others. Essential services, healthcare, basic consumer goods, and debt collection remain stable or grow during economic downturns. Luxury goods, discretionary services, and cyclical industries typically suffer more during recessions.

Consider building skills and income streams in areas that are recession resistant. This might include healthcare services, basic maintenance and repair, essential software services, or educational content. The key is identifying needs that persist regardless of economic conditions.

Build relationships and reputation in your chosen areas before you need them. During economic downturns, established relationships and proven track records become even more valuable. Clients are more likely to continue working with trusted providers and may be hesitant to try new vendors.

Long-Term Wealth Building

While preparing for uncertainty is essential, the ultimate goal is building substantial long-term wealth. This requires thinking beyond immediate needs and focusing on strategies that compound.

Compound interest remains one of the most powerful wealth-building forces available to individual investors. The key is to start early, invest consistently, and allow time for growth to accumulate. Even modest amounts invested regularly can grow into substantial sums over decades.

However, traditional buy-and-hold strategies may need modification in rapidly changing markets. While the principle of long-term investing remains sound, the specific implementation may need to be more active and adaptive.

Consider building wealth through ownership rather than just financial investments. This might include owning businesses, intellectual property, real estate, or other assets that can appreciate and generate income. Ownership provides more control over outcomes and can offer better protection against inflation and market volatility.

Tax optimization becomes increasingly essential as wealth grows. You can significantly affect long-term wealth accumulation by understanding how different income is taxed, how to time gains and losses, and how to use tax-advantaged accounts.

Estate planning provides an efficient and effective method for preserving and transferring wealth. This becomes important when wealth encompasses non-traditional assets, such as intellectual property, online businesses, or cryptocurrency.

Practical Implementation

Start by conducting a comprehensive financial audit. Document and list all income sources, expenses, assets, and liabilities. Identify which income sources are most vulnerable to technological disruption and which offer the best growth potential.

Set specific, measurable financial goals with clear timelines and deadlines. These include building emergency reserves, achieving certain investment returns, creating additional income streams, or reaching specific net worth targets. Having clear objectives makes it easier to track progress and make necessary adjustments.

Automate as much as possible. Automated saving and investing ensure consistent progress regardless of busy schedules or changing circumstances. Automated rebalancing maintains optimal asset allocation without requiring constant attention.

Continuously educate yourself about financial strategies, investment opportunities, and economic trends to stay informed and make informed decisions. The pace of change means that strategies that worked last year may not work next year. Staying informed enables you to adjust your approach as conditions develop.

Build relationships with financial professionals who understand the unique challenges of building wealth in rapidly changing times. This might include fee-only financial advisors, tax professionals who understand new economy tax issues, or investment advisors who specialize in the technology sector investments.

The Path Forward

Financial planning for an uncertain future requires accepting that traditional approaches may not be sufficient while also recognizing that fundamental principles, such as saving, investing, and diversifying, remain essential. The key is adapting these principles to new circumstances while capitalizing on new opportunities.

Success in this environment requires both defensive and offensive strategies. Defensive strategies protect against downside risks and provide stability during uncertain times. Offensive strategies position you to benefit from the opportunities that rapid change creates.

The goal isn't to predict the future perfectly, but to build financial resilience and positioning that can adapt to various scenarios. This means maintaining flexibility while also taking calculated risks that can generate substantial returns.

Remember that building wealth is a marathon, not a sprint. While the pace of change may speed up, successful wealth-building still requires patience, discipline, and consistent effort. The tools and strategies may evolve, but the underlying principles of spending less than you earn, investing the difference wisely, and staying committed to long-term goals remain constant.

The future belongs to those who prepare for it. By building diversified income streams, leveraging technology for wealth creation, and maintaining both flexibility and discipline in your financial approach, you can not only survive the coming changes but thrive in them. The question isn't whether change will come—it's whether you'll be ready when it does.

Chapter 16
Data Center Shortages, Power Demand, and Water Consumption

The rapid advancement of AI has transformed industries, from healthcare to finance, by enabling unprecedented computational capabilities. However, this progress has significantly increased the demand for data centers to support AI's intensive processing requirements. Training large language models and running complex algorithms necessitate vast amounts of computational power, which relies on robust data center infrastructure. As AI adoption accelerates, the global supply of data centers struggles to keep pace, leading to a critical shortage. The massive energy and water consumption required to operate these facilities compounds this bottleneck, raising concerns about sustainability and resource availability.

Data centers are the backbone of AI operations, housing servers that process and store enormous datasets. The scale of modern AI models, such as those powering generative AI tools, requires thousands of high-performance GPUs, which consume significant electricity. Reports from 2024 indicate that data centers in the United States accounted for approximately 4% of national electricity consumption, a figure projected to double by 2030 because of AI-driven demand. Companies like Microsoft and Google have announced multi-billion-dollar investments in new data centers, yet construction timelines lag the immediate needs of AI development, exacerbating the shortage. This rapid expansion underscores the urgent need for enhanced infrastructure to support AI's growth.

The energy-intensive nature of data centers has strained power grids worldwide, prompting innovative solutions and regional challenges. In Northern Virginia, a global hub for data centers, local governments have paused new projects because of insufficient power infrastructure, with utility provider Dominion Energy unable to guarantee supply. Similarly, in Ireland, data center expansion has stressed the national grid, leading to restrictions on new construction. Emerging markets, such as India and Southeast Asia, face similar constraints due to underdeveloped power grids. To address these challenges, companies are exploring alternative energy sources, with Microsoft planning to restart a decommissioned nuclear plant in 2025 to power its AI data centers, highlighting the scale of energy demands.

Beyond their immense energy requirements, data centers also have a substantial impact on water resources. These facilities rely heavily on water for cooling, using systems such as cooling towers and chillers to manage the heat generated by servers. The scale of this water consumption is significant; for instance, Google's hyperscale data centers use approximately 200 million gallons of water annually, while Microsoft reported a total water consumption of 1.69 billion gallons in a recent year.

A Bloomberg News analysis from May 2025 revealed that roughly two-thirds of new data centers built or in development in the U.S. since 2022 are located in areas with high levels of water stress, exacerbating local water scarcity issues. Globally, data centers currently consume 560 billion liters of water annually, with projections indicating a more than doubling to 1,200 billion liters by 2030.

In Virginia's "Data Center Alley," water consumption has surged by nearly two-thirds since 2019, reaching over seven billion liters in 2023. This high-water demand is particularly concerning in regions already facing water shortages, highlighting the need for sustainable water management practices in the data center industry.

However, this challenge also presents opportunities for innovation in water-efficient cooling technologies and sustainable water management practices. Companies and researchers are exploring alternative cooling methods, such as air cooling, liquid cooling, and even using wastewater or seawater, to reduce freshwater consumption. There is growing interest in water recycling and replenishment projects to offset the water used by data centers.

The shortage of data centers and the surging demand for power and water present significant opportunities for job creation across multiple sectors. The construction of new data centers requires skilled labor, including engineers, architects, and construction workers, to design and build state-of-the-art facilities. Electrical engineers and technicians are in high demand to develop and maintain the complex power systems that keep data centers operational. For example, Amazon Web Services has launched training programs to upskill workers for roles in data center operations, anticipating the need for thousands of new hires globally.

The push for sustainable energy and water solutions has created roles in renewable energy development, such as solar and wind farm technicians, and in water management, including specialists in cooling technology and water recycling. These roles are critical to addressing the environmental impact of data centers while supporting AI's expansion.

Beyond technical roles, the ripple effects of this demand extend to ancillary industries. Logistics and supply chain professionals are needed to manage the delivery of hardware components, such as GPUs and cooling systems, to data center sites. Electricians to install, maintain, and repair the massive electrical infrastructure needed to power high-performance computing equipment, including specialized high-voltage systems, backup generators, and cooling systems that keep AI servers running reliably. A growing need for policy experts and environmental consultants to navigate regulatory hurdles and ensure compliance with sustainability standards.

The global nature of this challenge also opens opportunities for international collaboration, with governments and corporations seeking experts to develop energy-efficient technologies and grid modernization strategies. For instance, initiatives like the European Union's Green Deal are funding projects that train workers in energy-efficient data center management, creating a pipeline of jobs. The focus on water conservation fosters opportunities for researchers and engineers to innovate in areas such as liquid cooling and wastewater reuse, further expanding the job market.

The rise of AI has exposed critical gaps in data center capacity, power availability, and water resources, driving a transformative shift in infrastructure development. While these challenges pose risks to AI scalability and environmental sustainability, they also create a wealth of opportunities for individuals seeking careers in technology, engineering, construction, and environmental management.

As companies and governments race to address these shortages, investments in workforce development, innovative energy solutions, and water-efficient technologies will be crucial. Professionals equipped with the skills to build, power, and maintain the next generation of data centers will find themselves at the forefront of a rapidly evolving industry, shaping the future of AI and global connectivity.

Part III: Action

Peter & Misty Phillip

Chapter 17
Your AI-Driven Career Path

If you've made it this far, you've done the hard work of understanding the landscape and crafting your strategy. You've identified where the opportunities lie, assessed your current position, and mapped out your path forward. Now comes the most crucial phase: execution.

The chapters that follow represent a fundamental shift in how we approach your upskilling journey. Where Parts I and II were designed for reflection, analysis, and strategic thinking, Part III is built for action. The pacing accelerates deliberately because in today's rapidly changing professional landscape, the gap between learning and doing must be as narrow as possible.

A New Format for a New Phase

You'll notice an immediate change in structure. Gone are the longer, narrative chapters that guided you through complex frameworks and deep analysis. Instead, now you'll find concise, list-based chapters that function as practical toolkits. Each chapter in this section is:

Immediately actionable: every framework can be implemented today.

Scannable and reference-ready: structured for quick consultation during real projects.

Results-focused: emphasizing outcomes over theory.

This is by design. When you're in execution mode, you need clarity, speed, and precision. You need to know exactly what to do next, not wade through extended explanations of why it matters.

Your Toolkit Awaits

Think of the following chapters as your professional survival kit. Some tools you'll reach for daily; others will serve you in specific situations. All bridge the gap between knowing what you need to do and actually doing it.

The frameworks ahead aren't meant to be read passively. They're meant to be used, adapted, and refined through practice. Keep this section close as you navigate the challenges and opportunities that lie ahead.

The time for planning is over. The time for doing has begun.

Navigating the Transition

Navigating the transition to an AI-enhanced career requires strategic thinking and deliberate action. The pace of change is coming so fast that it will be easy to fall behind. The key is to take small, consistent steps today. By taking action, this will help you build up your hybrid advantage.

The window for adaptation is narrowing rapidly as AI capabilities expand exponentially, making early action not just advantageous but essential for career survival. Organizations are already beginning to restructure roles around human-AI collaboration, and those who wait for the transformation to slow down will compete for increasingly scarce positions that don't require AI fluency.

The most successful professionals are those who recognize that this isn't a distant future scenario but a present reality demanding an immediate response. They understand that building the hybrid advantage isn't about making dramatic career pivots overnight, but about embedding AI awareness and complementary skills into their daily work routine. This might mean spending fifteen minutes each morning exploring how AI tools could streamline a routine task, dedicating one hour weekly to learning about AI applications in their industry, or volunteering for projects that involve human-AI collaboration.

The compound effect of these small, consistent actions creates a powerful momentum that separates adaptable professionals from those who become obsolete. Each incremental step builds upon the previous one, creating a foundation of AI literacy that enhances rather than replaces core competencies.

For instance, a marketing professional who starts by using AI for content ideation today will naturally progress to understanding customer behavior analytics, then to designing AI-enhanced customer experiences, ultimately positioning themselves as an invaluable bridge between traditional marketing wisdom and emerging AI capabilities. This progressive approach allows professionals to maintain their current productivity while systematically building future-ready skills. The hybrid advantage emerges not from abandoning proven expertise but from amplifying it with AI-powered capabilities, creating a professional profile that machines cannot replicate and competitors cannot easily match.

Here's a framework for identifying and pursuing opportunities:

Step 1: Inventory Your Assets

Begin by cataloging your existing knowledge, skills, and experience through an AI lens. Ask yourself:

What processes or problems do I understand deeply from my current or past work?

What human insights do I possess that AI systems might miss?

What relationships and networks have I built that could be valuable in new contexts?

What creative or analytical skills do I use that AI tools could enhance?

Remember, mundane experience often translates into valuable AI-era expertise. Years of customer service provide insights into human behavior that inform AI system design. Manufacturing experience offers an understanding of operational realities that AI implementation specialists need.

Step 2: Explore Adjacent Opportunities

Look for roles that combine your existing expertise with AI capabilities rather than abandoning your background entirely.

A few examples:

Finance professionals should: Explore AI-powered risk analysis, algorithmic trading oversight, or AI ethics in financial services. If you're in education, investigate AI tutoring system design, educational data analysis, or AI-powered curriculum development. Healthcare workers should consider AI diagnostic support, medical AI training, or AI-enhanced patient care coordination. As AI continues to develop the need for AI ethicists or digital trust officers will increase.

Beyond exploring these specialized AI applications, professionals across all industries must also embrace the transformative potential of AI agents to revolutionize their daily workflows. AI agents represent the next evolution in workplace automation, capable of handling complex, multi-step processes that previously required constant human oversight.

Finance professionals can deploy agents to monitor market conditions automatically, generate compliance reports, and flag unusual transaction patterns, while focusing on strategic decision-making and client relationships. These agents can work continuously in the background, processing vast amounts of financial data, updating risk models in real-time, and even initiating preliminary responses to market volatility based on predefined parameters. In education, AI agents can automate administrative tasks like grading, student progress tracking, and personalized learning path adjustments, freeing educators to concentrate on creative curriculum design and meaningful student interactions. Healthcare workers can leverage agents to streamline patient scheduling, medication management, and follow-up care coordination, while maintaining the human touch that remains essential in patient care.

Professionals implementing AI agents strategically must systematically consider which aspects of their work are automate-able and which demand human judgment and creativity. This shift demands a new skill set: the ability to design, train, and manage AI agents effectively. Professionals must learn to break down complex workflows into agent-manageable components, establish clear decision trees for automated responses, and create feedback loops that allow agents to improve.

As these technologies mature, emerging roles like AI Ethicist and Digital Trust Officer will become critical for ensuring that automated systems operate within ethical boundaries and maintain public confidence. These positions will require a deep understanding of both technical capabilities and human values, representing the ultimate expression of the hybrid advantage where technical proficiency meets ethical reasoning and stakeholder communication. The professionals who master agent deployment today will be positioned to lead the workforce transformation of tomorrow, serving as the bridge between human expertise and artificial intelligence capabilities.

Step 3: Identify Skill Gaps and Learning Paths

Once you've identified potential AI-enhanced career paths, assess what additional skills you need and create a learning plan.

Focus on:

Technical Literacy

You don't need to become a programmer, but understanding how AI systems work, their capabilities and limitations, and how to interact with them effectively is increasingly essential. Treat AI as a hyper-intelligent assistant, or a clone of yourself, that can accomplish many things you do, but faster and repeatedly.

Data Fluency

Many AI-era roles involve working with data in some capacity. Basic statistics, data visualization, and understanding of data quality become valuable across industries.

AI Tool Proficiency

Learn to use AI tools relevant to your field. This might include generative AI for content creation, AI analytics platforms, or industry-specific AI applications.

Human-AI Collaboration

Develop skills in working alongside AI systems, understanding when to trust AI recommendations and when to apply human judgment.

Step 4: Build and Demonstrate New Capabilities

Start applying AI tools and concepts in your current role to build evidence of your capabilities. Create projects that showcase your ability to combine domain expertise with AI technologies. Document your learning journey and results to show your adaptability and growth mindset.

Step 5: Network and Position Yourself

Connect with others in your target AI-enhanced field. Attend virtual conferences, join online communities, and engage with content creators in your area of interest. Share your learning journey and unique perspective on how AI might transform your industry.

Position yourself as someone who bridges traditional expertise with AI capabilities rather than competing with pure technologists. Emphasize your domain knowledge, ethical perspective, and human insights as complementary to AI technologies.

Chapter 18
AI Tools Upskilling Action Plan

Phase 1: Foundation Building

Understand the AI Landscape

Start by mapping the current AI ecosystem. Research the major categories of AI tools: generative AI (like ChatGPT, Claude), automation platforms (Zapier, Microsoft Power Automate), data analysis tools (Tableau with AI features), and industry-specific solutions. Read AI newsletters like The Rundown or AI Tool Report to understand trends and new releases.

Assess Your Current Needs

Audit your daily workflows to identify repetitive tasks, bottlenecks, or areas where you spend excessive time. Create a priority list of problems you'd like AI to solve, whether that's content creation, data analysis, customer service, design work, or process automation.

Phase 2: Hands-On Exploration

Try Core AI Tools

Set up accounts with major AI platforms and spend dedicated time learning their capabilities. For ChatGPT, Gemini, Grok, or Claude, practice different prompts for various use cases. Experiment with image generation tools like Midjourney or DALL-E if visual content applies to your work. Test automation tools by connecting simple workflows between apps you already use.

Explore Agentic AI

Begin exploring AI agents. Tools like Replit and n8n are revolutionizing the way we build and deploy AI-powered workflows, making them accessible even to those with minimal coding experience. Replit's AI Agent simplifies app development by turning natural language prompts into fully functional applications, handling everything from code generation to deployment in minutes.

Meanwhile, n8n's low-code, node-based platform empowers users to create sophisticated AI agents that integrate with hundreds of APIs, automating tasks like email processing, data analysis, or even complex multi-agent workflows. By exploring these tools, you can rapidly prototype and deploy intelligent systems, staying ahead in a world where automation and AI are reshaping industries.

Document Your Progress

Keep a learning journal tracking which tools you test, what you use them for, their strengths and limitations, and potential applications in your work. Note the pricing structures and any learning curves you encounter.

Phase 3: Practical Application

Implement Strategic Tools

Choose 2-3 AI tools that showed the most promise during your exploration phase and integrate them into your actual workflows. Start small with low-risk applications, then gradually expand usage as you become more comfortable and see results.

Measure Impact

Track time savings, quality improvements, or other relevant metrics from your AI tool usage. This helps justify continued investment and identifies which tools provide the most value for your specific situation.

Phase 4: Advanced Skills & Scaling

Develop Prompt Engineering Skills

Learn advanced techniques for getting better results from AI tools through more effective prompting. Study examples of successful prompts in your field and practice iterating on your requests to improve outputs. There are many ways to write prompts. You can assign a role to the AI, telling it to act as an expert in, say, marketing and direct its output, voice, look or impression. You can make the AI constantly prompt you for more feedback and go through as many iterations as needed to get exactly what you wanted.

Stay Current

Establish routines for staying updated on new AI tools and features. Follow relevant LinkedIn creators, join AI communities, and set up Google Alerts for AI tools in your industry. The AI landscape changes rapidly, so continuous learning is essential.

Key Resources to Explore

Discovery Platforms

Product Hunt's AI section for new tool launches (https://www. producthunt.com)

There's An AI For That database for comprehensive tool discovery

(https://theresanaiforthat.com)

AI tool comparison sites and review platforms

Use AI to help with this; it will keep you working with AI and help you develop a deeper understanding of AI and how to maneuver in the AI space.

Learning Communities

Reddit communities like r/artificial and r/ChatGPT

LinkedIn AI-focused groups in your industry

Discord servers focused on AI tools and automation

This plan gives you a structured approach to building AI fluency while ensuring you find tools that solve real problems in your work. The key is balancing exploration with practical application, so you develop both broad awareness and deep competency with the most valuable tools for your specific needs.

Chapter 19
Assessing Your Industry's AI Exposure

Overall Objective

Evaluate AI's impact on your industry and role, identify skill gaps, and begin upskilling to thrive in an AI-transformed workplace.

Week 1: Industry AI Exposure Assessment

Objective: Understand AI adoption and its impact on your industry and role.

Actions:

Research AI implementations in your sector (past 12-24 months) via industry reports (McKinsey, Deloitte) and 3-5 key publications.

Identify 3-5 leading companies using AI and map use cases (e.g., automation, data analysis).

Analyze your role's tasks, categorizing them by AI automation potential (High/Medium/Low).

Review five competitor job postings for AI-related roles.

Deliverable: Create a one-page report summarizing the AI landscape, competitor AI use, and personal role exposure.

Week 2: Future Trends and Skill Gap Analysis

Objective: Forecast AI's 12–24-month impact and identify required skills.

Actions:

Attend 1 webinar or review 2 white papers on AI trends in your industry.

Create a timeline of expected AI adoption milestones.

List current skills (technical/soft) and rate proficiency (Beginner/Intermediate/Advanced).

Research job descriptions for evolved versions of your role, noting skills in 80% + of postings.

Identify gaps in:

AI Literacy (capabilities, limitations)

Data Skills (analysis, visualization)

Human-AI Collaboration

Critical Thinking/Adaptability

Deliverable: AI transformation timeline and personal skill gap matrix.

Week 3: Strategic Upskilling—Tier 1 Skills

Objective: Build a foundational AI and data literacy.

Actions:

Enroll in "AI for Everyone" (Coursera/edX) and complete 50% of the course.

Practice prompt engineering with consumer AI tools (e.g., ChatGPT).

Learn basic Excel/Google Sheets functions for data analysis.

Experiment with one to two AI-enhanced productivity tools relevant to your role (e.g., Grammarly, Notion AI).

Allocate five hours to learning and hands-on practice.

Deliverable: Document progress in AI course and notes on tool usage.

Week 4: Implementation and Progress Review

Objective: Integrate AI tools into workflows and plan the next steps.

Actions:

Implement one AI tool in your daily workflow and document its impact.

Network with two to three professionals in AI-related roles via LinkedIn or industry forums.

Review progress in skill development and adjust goals.

Propose 1 AI pilot project or process improvement at work.

Outline a 3-month upskilling plan focusing on:

Complete an AI course and learn SQL basics.

Exploring industry-specific AI tools.

Building strategic thinking (e.g., AI ethics, business cases).

Deliverable: one-page summary of tool integration, networking outcomes, and three-month plan.

Resource Allocation

Time: five to seven hours/week for research, learning, and practice.

Budget: $50-100 for online course or free resources (e.g., Tableau Public, or Coursera.

Tools: access to ChatGPT, Excel/Google Sheets, and industry-specific AI platforms.

Success Metrics

Complete 50% of the AI course and implement one AI tool in the workflow.

Produce actionable deliverables (reports, matrices, plans).

Establish two to three professional connections in AI-related roles.

Gain confidence in discussing AI trends and their impact on your role.

Risk Mitigation

High AI Displacement Risk: prioritize human-AI collaboration and uniquely human skills (e.g., creativity).

Medium/Low Risk: focus on leveraging AI for productivity and exploring leadership opportunities.

Conclusion

This one-month plan equips you to assess AI's impact, begin upskilling, and position yourself as an AI-informed professional. Regular updates and consistent learning will ensure you stay ahead in an AI-driven workplace.

Chapter 20
Practical Upskilling Strategies for Mastering Technical Skills

In the time you were scrolling through social media last week, ChatGPT wrote a marketing campaign that landed a six-figure client. A data analyst at a Fortune 500 company got promoted because she automated her entire workflow with Python scripts. A customer service rep saved his position by learning to train AI chatbots instead of being replaced by them. Change is happening faster every day.

The robots aren't coming for your job; they're already here. This isn't science fiction. It's our current reality. By taking small, consistent steps everyday we can move forward.

The uncomfortable truth is every day you spend not learning these tools is a day someone else gains an advantage over you. The good news is you don't need a computer science degree to catch up. Start now to avoid falling behind.

Walk into any office, hospital, or retail store, and you'll see it happening. Marketing teams use AI to generate ad copy in minutes instead of days. Accountants let automation handle data entry while they focus on strategy. Even hairstylists use scheduling AI to manage their bookings.

The disruption isn't theoretical anymore.

It's happening in real-time, and every day it accelerates. Companies that embrace these tools gain massive competitive advantages. Workers who learn to use them become indispensable. Those who don't? Well, check the unemployment lines.

The AI revolution creates as many opportunities as it destroys. You need to position yourself now, so you are not left behind in the AI revolution.

The Three Skills You Need Now

Forget about becoming a coding expert overnight. Programs like Claude can generate code for you. Instead, focus on these three game-changing areas anyone can master. These skills build the foundation for the "human advantage" soft skills from Chapter 11.

1. Data Analysis

Learning data analysis is most powerful when combined with critical thinking and creative problem-solving to interpret that data in novel ways. Data analysis isn't about complex algorithms. It's about asking the right questions and finding answers. Every business decision now requires data backing, from hiring choices to marketing budgets. Companies today are drowning in tons of data but are starving for insights. The person who can turn numbers into actionable recommendations makes himself or herself invaluable.

Start here today:

Learn to create your first pivot table this week and identify one way to apply it to your work.

Learn to create charts that tell a story, not just display numbers.

Practice with your own data; track your expenses, fitness, or work hours.

Data Analytics Tools for Beginners

ChatGPT and Similar Large Language Models

A tool for conversational data analytics and code generation like ChatGPT. It writes Python or SQL scripts for data cleaning, analysis, or visualization and explains statistical concepts. Small business owners use it to generate scripts for calculating averages from CSV files.

Google Cloud AutoML

A tool for no-code custom machine learning like AutoML. It builds predictive models for classification or regression, analyzing structured data to predict outcomes like customer churn. Small businesses use it to forecast inventory needs from sales data.

Tableau with Built-in Intelligence

A tool for interactive data visualization and trend analysis like Tableau. It creates dashboards and uses machine learning to identify trends and outliers in data from Excel or databases. Marketing teams visualize campaign performance.

Microsoft Power BI

A tool for automated data insights and visualization like Power BI. It analyzes data from Excel or cloud sources, offering natural language queries and pattern detection for trends like employee turnover. HR managers use it to predict retention risks.

DataRobot

A tool for automated predictive modeling like DataRobot. It simplifies data preprocessing and model building for forecasting tasks like demand prediction. Retailers use it to forecast sales based on historical data.

Zapier for Data Workflows

A tool for no-code data workflow automation, like Zapier. It connects apps like Google Sheets and CRMs to automate data collection and integration with analytics tools. Freelancers use it to compile client data for analysis.

MonkeyLearn for Text Analysis

A tool for no-code text analysis like MonkeyLearn. It performs sentiment analysis, topic classification, and keyword extraction on text data like customer feedback. Social media managers analyze brand sentiment from comments.

RapidMiner

A tool for no-code data preparation and predictive modeling like RapidMiner. It automates data cleaning, visualization, and modeling for tasks like customer segmentation. Startups use it to analyze purchasing behavior.

Getting Started

Pick one tool that matches your specific needs - Tableau for visualization, ChatGPT or Claude for coding help, or Zapier for automation. Most tools offer free versions or trials that let you experiment without spending money. YouTube channels like freeCodeCamp and official documentation provide step-by-step guides. Practice with real data from your work or personal projects, like budget spreadsheets or sales reports, to make the learning process more relevant and practical.

2. AI Tool Mastery: Your New Superpower

You don't need to build AI, instead; you need to direct it. Think of AI tools as incredibly capable interns who never sleep, never complain, and work at lightning speed. Your job is learning to manage them effectively. While others debate whether AI is good or bad, smart workers are using it to 10x their productivity. They're getting promotions, starting businesses, and solving problems faster than ever before.

Start immediately:

Spend an hour with ChatGPT or Grok writing emails, brainstorming ideas, or summarizing reports.

Use Zapier to connect your apps; let it automatically save email attachments to Google Drive.

Try Canva's AI features to create presentations that look professionally designed.

Use AI to research topics, outline articles, and handle client communications. Don't let AI replace your skills; use AI to amplify them.

3. Basic Automation: Work Smarter, Not Harder

Automation isn't about robots on factory floors; it's about eliminating the repetitive tasks that eat your day. Every manual process you automate frees up time for high-value work. Stop doing robot work so you can focus on human work. While others spend hours on data entry, you can build relationships and solve complex problems.

Quick wins to try this week:

Use Python to rename hundreds of files instantly (there are copy-paste scripts online).

Set up email filters to automatically organize your inbox.

Create simple scripts to generate reports instead of building them manually each time.

Learning Resources

Stop saying you can't afford to learn. Here's how to upskill without breaking the bank:

Free and Immediate:

Use YouTube: search "Excel for beginners" or "Python automation" and start watching videos.

There are thousands of hours of coding lessons that are completely free at freeCodeCamp.org.

Experiment daily with ChatGPT, Claude, Grok, and Gemini to master their capabilities. Explore the strengths and nuances of each model and refine your prompting techniques. These AI tools are your learning partners for effective interaction.

Worth the Investment:

Udemy courses: $10-$50 during sales (which happen constantly).

Coursera certificates: $39-$79 monthly, cancel anytime.

Community college courses: $200-$500 for semester-long programs.

Local and Hands-On:

Meetup.com: Find local coding, data analysis groups or user groups.

Library workshops: Many libraries offer free tech classes.

Community centers: look for digital literacy programs.

Don't overthink this. Choose one resource and begin today. Analysis paralysis is just procrastination with a fancy name.

Chapter 21
The 30-Day AI Survival
Challenge

Stop planning and start doing. This 30-day challenge will give you practical skills that immediately affect your work:

Week 1: Foundation Building

Days 1-3: Choose your focus area (data analysis, AI tools, or automation). Download the software. Complete your first tutorial.

Days 4-7: Practice daily for 45 minutes. Create something tangible, a chart, an automated email, or a simple script. Document your progress.

Week 2: Real-World Application

Days 8-14: Apply your new skill to an actual work problem. Analyze your department's data, automate a repetitive task, or use AI to improve a process. Discuss your results with a colleague or your boss.

Week 3: Expansion and Connection

Days 15-21: Take on a bigger project. Join an online community. Share your work on LinkedIn. Get feedback and iterate on your solutions.

Week 4: Portfolio Building

Identify a potential bias in an AI tool you use or draft a personal ethical guideline for using generative AI.

Days 22-30: Polish your best work. Create a simple portfolio showcasing what you've built. Plan your next learning goal. Schedule it on your calendar. Hold yourself accountable.

Success Strategies That Actually Work

Make it social: join online communities. Share your progress with others. Ask questions. The people learning alongside you today might hire you tomorrow.

Focus on problems, not tools: Don't learn Python for Python's sake. Learn it to solve a specific problem you're facing. Motivation will carry you through the hard parts.

Celebrate small wins: Automated your first email? Celebrate. Created your first chart? Share it on social media. Build confidence with minor victories before tackling bigger challenges.

Apply immediately: Don't wait until you're "ready." Use your half-formed skills on actual problems. You'll learn faster through doing than through studying.

The Harsh Reality

While you're reading this, someone else is learning these skills. They're automating their work, impressing their managers, and positioning themselves for the jobs of the future. Every day you delay is a day they pull further ahead.

The AI revolution is here. Companies are already deciding about who stays and who goes based on who can adapt to this new reality. The workers who learn to collaborate with AI will thrive. Those who don't will struggle.

But here's the thing about revolutions: they create opportunities for those bold enough to seize them. The same tools disrupting industries are available to you right now, often for free.

It is time to move. The next 30 days will determine whether you're riding the wave or getting swept away by it.

Start Today, Not Tomorrow. Close this book. Open YouTube. Search for one tutorial related to data analysis, AI tools, or basic automation. Watch it. Then do what it teaches.

Your future self, the one with job security, better opportunities, and higher income, is waiting for you to take that first step...

End Notes

Briggs, Joseph, and Devesh Kodnani. "The Potentially Large Effects of Artificial Intelligence on Economic Growth." Goldman Sachs Global Investment Research, March 26, 2023. https://www.goldmansachs.com/insights/articles/generative-ai-could-raise-global-gdp-by-7-percent

Chui, Michael, Eric Hazan, Roger Roberts, Alex Singla, Kate Smaje, Alex Sukharevsky, Lareina Yee, and Rodney Zemmel. "The Economic Potential of Generative AI: The Next Productivity Frontier." McKinsey & Company, June 14, 2023. https://www.mckinsey.com/capabilities/mckinsey-digital/our-insights/the-economic-potential-of-generative-ai-the-next-productivity-frontier

Electric Power Research Institute. "Powering Intelligence: Analyzing Artificial Intelligence and Data Center Energy Consumption." May 2024. https://www.epri.com/research/products/000000003002028786

Hinton, Geoffrey. "Will Digital Intelligence Replace Biological Intelligence?" Presentation at EmTech Digital, MIT Technology Review, May 2, 2023. https://www.youtube.com/watch?v=sitHS6UDMJc

International Energy Agency. "Electricity 2024: Analysis and Forecast to 2026." January 2024. https://www.iea.org/reports/electricity-2024

Kahn, Brian. "How AI Demand Is Draining Local Water Supplies." Bloomberg, May 8, 2025 https://www.bloomberg.com/graphics/2025-ai-impacts-data-centers-water-data/

Musk, Elon. Interview at MIT Aeronautics and Astronautics Centennial Symposium, October 24, 2014. https://www.youtube.com/watch?v=Tzb_CSRO-0g

OpenAI. Testimony of Sam Altman before the U.S. Senate Judiciary Subcommittee on Privacy, Technology, and the Law, May 16, 2023. Transcript and coverage in *New York Times* https://www.nytimes.com/2023/05/16/technology/openai-altman-artificial-intelligence-regulation.html

U.S. Department of Energy. "Assessment of Electricity Demand Growth from Electrification, AI, and Data Centers." December 20, 2024. https://www.energy.gov/articles/doe-releases-new-report-evaluating-increase-electricity-demand-data-centers

Williams, Aime. "US Tech Groups' Water Consumption Soars in 'Data Centre Alley'." *Financial Times*, August 17, 2024. https://www.ft.com/content/1d468bd2-6712-4cdd-ac71-21e0ace2d048

World Economic Forum. "The Future of Jobs Report 2020." October 2020. https://www.weforum.org/reports/the-future-of-jobs-report-2020

www.ingramcontent.com/pod-product-compliance
Lightning Source LLC
Chambersburg PA
CBHW070929210326
41520CB00021B/6855